Soil hydrological impacts and climatic controls of land use and land cover changes in the Upper Blue Nile (Abay) basin

SOIL HYDROLOGICAL IMPACTS AND CLIMATIC CONTROLS OF LAND USE AND LAND COVER CHANGES IN THE UPPER BLUE NILE (ABAY) BASIN

DISSERTATION

Submitted in fulfillment of the requirements of

the Board for Doctorates of Delft University of Technology

and
of the Academic Board of the UNESCO-IHE

Institute for Water Education

for
the Degree of DOCTOR

to be defended in public on

Tuesday, 7 July 2015, at 15:00 hours

In Delft, the Netherlands

by

Ermias Teferi DEMESSIE

Master of Science of Earth Science, Addis Ababa University

born in Wolieso, Ethiopia.

This dissertation has been approved by the
promotor: Prof. dr. S. Uhlenbrook

Composition of the doctoral committee:

Chairman	Rector Magnificus TU Delft
Vice-Chairman	Rector UNESCO-IHE
Prof.dr. S. Uhlenbrook	UNESCO-IHE/ TU Delft, promotor

Independent members:	
Prof.dr. J. Nyssen,	Ghent University, Belgium
Dr. B. Simane,	Addis Ababa University, Ethiopia
Prof.dr. W.G.M. Bastinaassen,	UNESCO-IHE/ TU Delft
Prof.dr. K.A. Irvine,	UNESCO-IHE / Wageningen University
Prof.dr. D. P. Solomatine,	UNESCO-IHE/ TU Delft
Prof.dr.ir. M.E. McClain	TU Delft/ UNESCO-IHE, reserve member

CRC Press/Balkema is an imprint of the Taylor & Francis Group, an informa business

© 2015, Ermias Teferi Demessie

Published by:
CRC Press/Balkema
PO Box 11320, 2301 EH Leiden, The Netherlands
e-mail: Pub.NL@taylorandfrancis.com
www.crcpress.com – www.taylorandfrancis.com

ISBN: 978-1-138-02874-6 (Taylor & Francis Group)

ABSTRACT

Land use and land cover change (LULCC) is a locally pervasive and globally significant environmental trend and has become a process of paramount importance to the study of global environmental change. LULCC is a widespread phenomenon and has contributed to the existing high rate of soil erosion and land degradation in the highlands of Ethiopia. Specifically, it has significant eco-hydrological impacts on soil physical and hydrological properties.. This thesis analyses LULCC and its links to soil hydrology, soil degradation and climate variables in the Upper Blue Nile (UBN) river basin of Ethiopia.

It is neither feasible nor realistic to carry out a detailed long-term LULCC investigation at the basin scale. The approach followed in this thesis was to take a case study site located in one of the watersheds of the source region of UBN river. Therefore, the long-term LULCC was quantified and the spatial determinants of land cover transitions were identified for the period 1957 to 2009 in the Jedeb watershed (296.6 km^2). Black and white aerial photographs of 1957 and Landsat imageries of 1972 (MSS), 1986 (TM), 1994 (TM) and 2009 (TM) were used to derive ten land use and land cover classes by integrated use of Remote Sensing (RS) and Geographic Information System techniques. The results showed that 46% of the study area experienced a transition over the past 52 years, out of which 20% was due to a net change while 26% was attributable to swap change (i.e. simultaneous gain and loss of a given category during a certain period). The most systematic transitions are conversion of grassland to cultivated land (14.8%) followed by the degradation of natural woody vegetation and marshland to grassland (3.9%). Spatially explicit logistic regression modelling revealed that the location of these systematic transitions can be explained by a combination of accessibility, biophysical and demographic factors. The modelling approach allowed improved understanding of the processes of LULCC and for identifying explanatory factors for further in-depth analysis as well as for practical interventions for watershed planning and management.

Conversion of wetlands for agricultural use is a type of LULCC that is widespread in Ethiopia. However, the effects of this conversion are poorly documented. Thus, wetland loss was quantified using satellite imageries of the period 1986-2005. The results showed that 607 km^2 of seasonal wetland with low moisture and 22.4 km^2 of open water areas are lost in the study area. The current situation in the wetlands of

Choke Mountain is characterized by further degradation, which urgently calls for wetland conservation and rehabilitation efforts through incorporating wetlands into watershed management plans.

The enhanced transition matrix provides useful information about change on land cover proportions, but it does not show the composition and configuration of the landscape that can be revealed through landscape pattern metrics. Different landscape metrics indicated that the spatio-temporal pattern of the Jedeb landscape has undergone significant alterations in terms of composition and configuration since 1957. Landscape fragmentation has increased during the period 1957 to 2009 as discerned by the increase in the number of smaller patches (an increase in the number of patches (NP) from 1621 to 5179 and a decrease in the mean patch size from 18.3 to 5.7 ha). The relative variability of the patch size increased over time from 1743 to 4933 ha suggesting that landscape heterogeneity increased posing a land management challenge. In the year 2009 the Jedeb watershed was dominated by a few contiguous patches and all patch types were aggregated as the contagion index (CONTAG) has increased from 57% to 65% during the period 1957-2009. This increase in dominance is also supported by the increase in the largest patch index (LPI) from 42% to 68% and a decrease in the Shanon's Diversity index (SHDI) value from 1.47 to 1.11 in 1957 and 2009, respectively. The mean shape index (MSI) ranked the Jedeb landscape along an intuitive temporal gradient from most complex to simple shapes (1.57 to 1.24). The patches were most irregular and complex in shape during the year 1957, whereas patches became less irregular and simpler in shape progressively from 1957 onwards. These results indicate that fragmentation is causing a simplification in patch shape compared to the geometrically complex patch shapes found in the past.

Soil degradation is one form of degradation that is expected to be caused by LULCC. This thesis evaluated the influence of LULCC on key soil quality indicators (SQIs) in the Jedeb watershed of the Blue Nile/Abay Basin. A factorial (2x5) multivariate analysis of variance (MANOVA) showed that LULC and altitude together significantly affected soil organic matter levels. However, LULC alone significantly affected bulk density and altitude alone significantly affected bulk density, soil acidity, and silt content. Afforestation of barren land with *Eucalyptus* trees can significantly increase the soil organic matter in the midland part but not in the upland part. Soils under grassland had a significantly higher bulk density than did soils under natural woody vegetation indicating that de-vegetation and conversion to grassland could leads to soil compaction. Thus, the historical LULCC in the Jedeb watershed has resulted in the loss of soil organic matter and increased soil compaction. The study shows that a land use and management practice can be monitored if it degrades, maintains, or improves the soil using key soil quality indicators.

LULCC can cause changes in soil physical properties and total evaporation to result in changed soil moisture dynamics. In this thesis, both the downscaled and *in-situ* soil moisture observations were used to evaluate the impact of land cover change on soil moisture dynamics at basin (199,812 km^2) and watershed (296.6 km^2) scales. Results of remote sensing observation indicated that there are significant differences in the mean soil moisture content between the reference trajectories (stable forest) and other trajectories considered as determined by the ANOVA result ($p < 0.001$). Soils under stable forest trajectories had the highest soil moisture level. *In-situ* observations from the case study watershed (Jedeb) also confirmed that the mean soil moisture levels under barren land and grassland significantly differed from the mean soil moisture content of forest soils ($p < 0.001$). Thus, it could be demonstrated that land cover changes have a significant influence on soil moisture regime across all scales at the basin.

Measurements on the full range of soil moisture retention characteristics curve are often not available for use in hydrological models to assess land use impacts on soil hydraulic properties. Thus, soil water retention characteristics are often predicted from readily available soil data using pedotransfer functions (PTFs). Due to distinctive pedological properties of high altitude tropical soils, PTFs developed elsewhere cannot be applied to the soils of the Upper Blue Nile/Abay region. Basic soil physical and chemical properties were found to predict the soil water retention characteristics reasonably well with a mean root mean squared difference (MRMSD) of 0.0349 cm^{3}cm^{-3} and 0.0508 cm^{3}cm^{-3} for point PTFs and the continuous PTFs, respectively. Collecting and integrating fragmented available soil survey data enables scientists and planners to generate new information such as maps of available water and soil water retention characteristics at the basin scale. This study also suggested that future studies are recommended to be carried out using a data base that is large enough to be representative of almost all parts of the textural triangle.

Seasonal and inter-annual changes in vegetation condition are not accounted for in the broad time-scale LULCC. However, changes that might be of crucial importance could be obscured and seasonal and inter-annual changes in the vegetated land cover may be much greater than that resulting from the broad time-scale LULCC. Thus, this thesis examines inter-annual and seasonal trends of vegetation cover in the Upper Blue Nile/ Abay (UBN) basin. Advanced Very High Resolution Radiometer (AVHRR) based Global Inventory, Monitoring, and Modelling Studies (GIMMS) Normalized Difference Vegetation Index (NDVI) was used for coarse scale long-term vegetation trend analysis. Moderate-resolution Imaging Spectroradiometer (MODIS) NDVI data (MOD13Q1) was used for finer scale vegetation trend analysis. Harmonic analyses and non-parametric trend tests were applied to both GIMMS NDVI (1981–2006) and MODIS NDVI (2001-2011) data sets. Based on a robust trend estimator (Theil-Sen slope) most part of the UBN (~77%) showed a positive trend in monthly GIMMS NDVI with a mean rate of 0.0015 NDVI units (3.77% yr^{-1}), out of which 41.15% of the basin depicted significant increases ($P < 0.05$) with a mean rate of 0.0023 NDVI units (5.59% yr^{-1}) during the period. However, the finer scale (250 m) MODIS-based vegetation trend analysis revealed that about 36% of the UBN shows a significantly decreasing trend ($P< 0.05$) over the period 2001-2011 at an average rate of 0.0768 NDVI yr^{-1}. Seasonal trend analysis was found to be very useful in identifying changes in vegetation condition that could be masked if only inter-annual vegetation trend analysis was performed.

Attribution of the observed vegetation trend to controlling climatic factors is crucial. Thus, time series analysis of net primary productivity (NPP, amount of atmospheric carbon fixed) and water-use efficiency (WUE, amount of carbon uptake per unit of water use) and correlation analyses were conducted. The results show that the dominant climatic controlling factors of NPP vary according to aridity/ humidity classes as follows: rainfall and temperature in humid zones; temperature and vapour pressure deficit (VPD) in semi-arid zones; and cloudiness in dry sub-humid zones of the basin. In the dry sub-humid and humid zones of the basin, WUE was correlated significantly and positively with maximum temperature, potential evaporation, and VPD, although no single climatic factor was correlated with WUE in semi-arid zones of the basin.

To conclude, this thesis showed that understanding of LULCC and its link to land degradation, soil hydrology, biodiversity (landscape fragmentation), and climatic varibles is crucial in order to better manage its potential effects. Future studies should include the impact of LULCC on local and regional climate, the underlying forces of LULCC from the social science perspective, and the soil hydrological impacts of LULCC using both laboratory and field experiments. Moreover, the climatic controls of NPP and WUE at different agro-ecological zones of the Abay basin need to be studied further using higher spatial resolution satellite imageries and field observation in future studies.

ACKNOWLEDGEMENTS

First and foremost I would like to thank God. Without His grace and mercy, I wouldn't be living in this world today. Almighty, You have given me the power to believe in my passion and pursue my dreams. I could never have done this without the faith I have in You. Glory Be to Your Holy Name!!!

I respectfully and gratefully acknowledge my promoter Prof. Dr. habil. Stefan Uhlenbrook for accepting me as his student and for his insightful scientific guidance throughout the research project. Stefan, your critical comments and suggestions on the thesis, including field visits to my research area in Ethiopia have considerably helped to own its present state. My ideas have evolved over the years and been shaped by fruitful and often inspiring discussions with you and my supervisors. Stefan, by working in your group I also learnt many things, besides the PhD, which will be of great help for my future career.

I would like to extend my sincere thanks to my supervisors: Dr. Woldeamlak Bewket, Dr. Belay Simane and Dr. Jochen Wenninger for longstanding collegial relationships, the sharing of their ideas that often challenged my thinking, and their invaluable reviews and critiques over the years. I owe many thanks to Dr. Woldeamlak Bewket for his consistent encouragement to pursue excellence in every component of this study. Thank you for the confidence that you have placed in me and for the social discussions that we had together. A profound gratitude goes to Dr. Jochen Wenninger, who has been a truly dedicated supervisor when I was in The Netherlands. My sincere gratitude also goes to Dr. Belay for his overall contribution in this research, especially for his considerate response to all the professional and administrative issues, which greatly helped to the successful completion of this study.
I would like to express my sincere gratitude to the Centre for Environmental and Developmental Studies in the College of Development Studies, Addis Ababa University, for giving me the study leave. I gratefully acknowledge the funding sources that made my PhD research possible. The study was carried out as a project within a larger research program called "In search of sustainable catchments and basin-wide solidarities in the Blue Nile River Basin", which was funded by the Foundation for the Advancement of Tropical Research (WOTRO) of the Netherlands Organization for Scientific Research (NWO), UNESCO-IHE and Addis Ababa University. I am grateful to the Netherlands Fellowships Programme (NFP) for providing financial

x

support during the last three years of my PhD. I would further like to acknowledge the National Centre of Competence in Research (NCCR) North-South program for providing training pertaining to this thesis.

I would be remiss not to acknowledge the invaluable contributions of the Blue Nile Hydrosolidarity project director, Prof. Pieter van der Zaag, whose leadership and insights were instrumental in creating solutions to many of the difficult problems encountered in the whole project. The support made by other members of the project team is gratefully acknowledged with deep appreciation and with apologies for any misrepresentations. Although it is impossible to mention everyone at the project who contributed ideas and assisted me with comments and feedback, Dr. Melesse Temesgen, Sirak Tekleab, Rahel Muche, Hermen Smit, Dr. Yasir Saleh and Dr. Abonesh have been regular collaborators as well as supportive friends.

This thesis benefited tremendously from valuable critical comments provided by numerous colleagues from the UNESCO-IHE community. Among the most memorable of these are discussions with Dr. Girma Yimer, Patricia Trambauer, Yared Bayissa, Fikadu Worku, Yenesew Mengiste and Adey Nigatu.

Words cannot express how grateful I am to my wife, Ts'ge Addis, for all innumerable sacrifices that you've made on my behalf, especially during my absence in the course of producing this thesis. My daughter Abigya Ermias also deserves my appreciation for her patience while I have been busy pursuing my research. Abigu, you are the best gift that I have ever been given in my life! You welcomed me into fatherhood and I am so grateful for you. Daddy loves you more than you will ever know. Last but not least, I wish to express my heart-felt gratitude and appreciation to my parents Ato Teferi Demessie and Wzo Felekech T/Wold who have been there for me from day one of my life. This is a tribute to the four of you.

Ermias Teferi Demessie
Delft, The Netherlands
June 2015

TABLE OF CONTENTS

LIST OF SYMBOLS

Symbol	Parameter description	dimension
A	Surface albedo	
A^*	Scaled surface albedo	
α_0	Mean of the harmonic series	
α	Shape parameter of the soil moisture retention curve	
φ_n	Phase angle	
β_n	Regression coefficients	
B	Transmissivity factor	LT^{-1}
C_n	Amplitude	$ML^{-1}T^{-2}$
C_j	The total change for each land class	
D_j	The difference between gain and loss	
E	Total evaporation	LT^{-1}
E_a	Actual evaporation	LT^{-1}
E_p	Potential evaporation	LT^{-1}
e	Error term	
$f(t)$	Fourier series value at time t	
F_r	Fractional vegetation cover	
G_{ij}	Expected gains	
g_p	Gain-to-persistence ratio	
$\partial h/\partial z$	Change in soil water matric potential head over distance	
$h(\theta)$	Soil water matric potential head as a function of moisture content	L
$i(t)$	Infiltration rate at time t	LT^{-1}
$K(\theta)$	Hydraulic conductivity as a function of moisture content	LT^{-1}
k	Kappa Coefficient	
l_p	Loss-to-persistence ratio	
L_{ij}	Expected losses	
$LR\,\chi^2$	Likelihood Ratio or Model Chi-square	
M_0	Soil surface moisture availability	
M	Disaggregated microwave soil moisture	

m	Shape parameter of the soil moisture retention curve	
n	Shape parameter of the soil moisture retention curve	
$N*$	Scaled vegetation index	
p	Probability value	
P_{i+}	the proportion of the landscape in category i in time 1	
P_{+j}	The proportion of the landscape in category j in time 2	
P_{ij}	The proportion of the land that experiences transition	
P	Precipitation	LT^{-1}
pF	The logarithm to the base 10 of the soil moisture tension	
q	Soil water flow	L^{-3}
$\partial\theta/\partial t$	Change in soil moisture over time	LT^{-1}
θ	Volumetric water content	
θ_{FC}	Soil water content at field capacity	
θ_{WP}	Soil water content at permanent wilting point	
$\theta_{m,i}$	Measured values at time/place i	
$\theta_{p,i}$	Predicted values at time/place i	
$\bar{\theta}$	Average value of the measured data	
θ_r	Residual water content	
θ_s	saturated water content	
r	The annual rate of change	
r_s	Spearman's rank correlation coefficient	
R_{ij}	The ratio of gain to loss	
S_j	The amount of swap of land class j	
T	The length of the Fourier series	T
T_{dew}	Dew point temperature	Θ
S_t	Seasonal component	
T_t	Trend component	
T_s	Land surface temperature	Θ
T_{min}	Land surface temperature pertaining to a dense vegetation	Θ
T_{max}	Land surface temperature pertaining to a dry bare soil	Θ
$T*$	Scaled land surface temperature	
μ	Mean	

ω_n	Fourier frequency
x	The original value of the independent variable
Y_t	Observed NDVI data
σ	Standard deviation
z	Distance in the direction of flow
Z	Standardized variable

L

LIST OF ACRONYMS

A	Albedo
AD	Anderson-Darling
AEZ	Agroecological Zones
AIC	Akaike Information Criterion
AMSR-E	Advanced Microwave Scanning Radiometer-Earth Observing System
ANN	Artificial Neural Network
ANOVA	Analysis of Variance
ASCAT	Advanced SCATterometer
ASCAT	Advanced Scatterometer (Metop)
ASTER	Advanced Space-borne Thermal Emission and Reflection Radiometer
AVHRR	Advanced Very High Resolution Radiometer
AWC	Available Water Content
AWMSI	Area-weighted mean shape index
BFAST	Breaks For Additive Seasonal and Trend
BIC	Bayes Information Criterion
BMP	Best Management Practices
CASA	Carnegie-Ames-Stanford approach
CATDS	Centre Aval de Traitement des Donn´ees SMOS
CDTI	Centro para el Desarrollo Tecnologico Industrial
CLUMPY	Clumpiness index
CMK	Contextual Mann Kendall
CNES	Centre National d'Etudes Spatiales
CONNECT	Connectance index
CONTAG	Contagion
CRU	Climate Research Unit
CSA	Central Statistical Agency
DEM	Digital Elevation Model
DOY	Day of Year
DTED	Digital Elevation Terrain Data
ECV	Essential Climate Variable
ED	Edge density
EMA	Ethiopian Mapping Agency

ENN_MN	Mean Euclidean Nearest Neighbor Distance
EROS	Center for Earth Resources Observation and Science
ERS	Europian Remote Sensing Sate
EVI	Enhanced Vegetation Index
FC	Field Capacity
Fr	Fractional vegetation cover
GCP	Ground Control Point
GIMMS	Global Inventory, Monitoring, and Modelling Studies
GIS	Geographic Information System
GLADA	Global Assessment of Land Degradation and Improvement
GLO-PEM	Global Production Efficiency Model
GPCC	Global Precipitation Climatology Centre
HANTS	Harmonic ANalysis of Time Series
HYPRES	Hydraulic Properties of European Soils
IRS	Indian Remote Sensing satellites
ISODATA	Iterative Self-Organizing Data Analysis
KBP	Koenker's studentized Bruesch-Pagan
k-NN	k-nearest neighbour
LPI	Largest patch index
LULCC	Land use and land cover change
LUPRD	Land use Planning and Regulatory Department
MANOVA	Multivariate Analysis of Variance
MLP	Multilayer Perceptron
MODIS	Moderate-resolution Imaging Spectroradiometer
MODIS	Moderate-resolution Imaging Spectroradiometer
MPS	Mean patch size
MRMSD	mean root mean squared difference
MSI	Mean shape index
MSIDI	Simpson's Diversity Index
MSIEI	Shannon's evenness index
MSS	Multispectral Scanner
MVC	Maximum Value Composite
NDVI	Normalized Difference Vegetation Index
NP	Number of patches
NPP	Net Primary Productivity
OLS	Ordinary Least Square
PCA	Principal Component Analysis
PLAND	Percentage of landscape
PROX_MN	Mean proximity index
PSCV	Patch size coefficient of variation
PSD	Particle Size Distribution
PWP	Permanent Wilting Point
ROC	Relative Operating Characteristics

RS	Remote Sensing
SAR	Synthetic Aperture Radar
SLM	Sustainable Land Management
SMMR	Scanning Multichannel Microwave Radiometer
SMOS	Soil Moisture Ocean Salinity
SMU	Soil Mapping Units
SPOT	Satellite Pour l'Observation de la Terre
SQI	Soil Quality Indicators
SSM/I	Special Sensor Microwave/Imager
SVAT	Soil Vegetation Atmosphere Transfer
SWRC	Soil Water Retention Characteristics
TDR	Time Domain Reflectometry
TM	Thematic Mapper
TMI	Tropical Rainfall Measuring Mission Microwave Imager
TMN	monthly average daily minimum temperature
TMP	monthly average daily mean temperature
TMX	monthly average daily maximum temperature
TOA	Top-Of-Atmosphere
TPP	Trend Preserving Prewhitening
Ts	Land Surface Temperature
TWI	Topographic Wetness Index
UBN	Upper Blue Nile
UNSODA	Unsaturated Soil Hydraulic Database
URMSR	Unbiased Root Mean Square Residual
USGS	United States Geological Survey
UTM	Universal Transverse Mercator
VI	Vegetation Index
VPD	Vapor Pressure Deficit
VWC	Volumetric Water Content
WPISPP	Woody Biomass Inventory and Strategic Planning Project
WUE	Water Use Efficiency

Chapter 1

INTRODUCTION

1.1 BACKGROUND

In the second half of the 20th century, humans have degraded land resources more rapidly than in any comparable period of time in the history of humankind, mainly to meet increasing demands for food, water and energy as a result of growing population (Godfray et al., 2010; MEA, 2005; Nyssen et al., 2004). As the demand for food increases resulting from the growth in population and *per capita* consumption (Kearney, 2010), vulnerability to food insecurity increases as a result of increased competition for land and water (De Fraiture et al., 2008; Savenije and Van Der Zaag, 2002). Moreover, food producers become more stimulated to bring more new land into large-scale commercial farming through deforestation and drainage. Thus, feeding many more people while keeping the environmental integrity in an increasingly uncertain economic situation is one of the major challenges the world is facing in the first quarter of the 21st century (Godfray et al., 2010; Kearney, 2010; McClain, 2013). According to Global Assessment of Land degradation and Improvement (GLADA) project, land degradation affects nearly 24% of the Earth's land area and has resulted in the loss of Net Primary Productivity (NPP) of 956 metric ton (van den Berg et al.) of Carbon relative to the period 1981–2003. Globally, nearly 47%, 25% and 18% of the degrading areas are associated with forest, grassland, and agricultural land, respectively (Bai et al., 2008).

Ethiopia's population, which is the second largest in sub-Saharan Africa after Nigeria's, is growing fast. Ethiopia also has the largest livestock population in Africa. This has created an increased pressure on the limited and fragile land resources. The demand for food, fuel, timber, forage and browse without effective sustainable management is leading to an increasing depletion of the country's natural vegetation, particularly from pastures, woodlands and forests (Nyssen et al., 2004; Teferi, 2011).

More than 90% of Ethiopia's population lives in the highlands (above 1500 m a.s.l), where there is also about 93% of the cultivated land, around 75% of the country's livestock and over 90% of the country's economic activity. Consequently, these highlands are significantly affected by land degradation. About 57%, 8% and 20% of Ethiopia's land area is moderately, severely, or very severely degraded, respectively (Bot et al., 2000). Thus, land degradation is seriously threatening the economic and social development of the whole country (Sonneveld, 2002).

The severity of droughts and the resulting famines in 1972/73 and 1984/85 can be partly attributed to an accelerating process of land degradation combined with widespread poverty of the growing population (Bisrat, 1990). The succession of droughts over the past 20 years and severe deficit of animal feed have led to more deterioration in the rangelands. The demand for cropland has been increasingly competing with that for rangeland. The long-term dynamics of rangelands in relation to other land-uses are still inadequately understood. In certain areas, the situation has been exacerbated by the loss of rangelands to small/large-scale irrigation and to encroachment by cultivators, placing increased demands on the remaining rangeland resources (Bisrat, 1990; Teferi et al., 2013). In Ethiopia, grazing and browsing occur in more than 50% of the country's land area and grazing lands are not owned by individuals or groups of people and this leads to a situation that more or less fits to the "tragedy of the commons", where grazing lands are exploited well above their carrying capacities. This heavy pressure constitutes a huge threat to the environment and society.

1.2   STATE OF THE ART

1.2.1   Soil Hydrological Processes

The interaction of the soil with the various components of the hydrological cycle is central for the dynamics of hydrological cycle and is known as soil hydrology. Soil hydrology has a strong influence on the water uptake and release by plants during photosynthesis (Shukla, 2011). Soil hydrology takes into account all of the components of water related to infiltration, root and plant water uptake and release, evaporation from soil and plant surfaces, transpiration, lateral flows, and percolation and recharge to groundwater (Fig. 1).

Soil water is held in an unsaturated zone by forces whose effect is expressed in terms of the energy state, pressure or potential of the water. Of the various components of this potential, the matric potential manifests the tenacity with which soil water is held by the soil matrix. The direction and amount of soil water flow between two locations in the soil profile can be determined from the total soil-water potentials at these locations. Soil water always moves in the direction of decreasing potential.

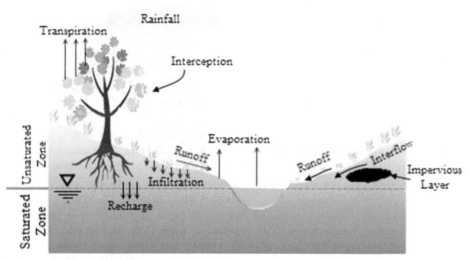

Figure 1 The water balance of a root zone

The amount of soil water flow (q) for unsaturated media is expressed by Darcy-Buckingham law (Buckingham, 1907) given by Eq. 1.1.

$$q = -K(\theta)\left(\left(\frac{\partial h(\theta)}{\partial z}\right) - 1\right)$$ [1.1]

Where $K(\theta)$ is unsaturated hydraulic conductivity $h(\theta)$ is often referred to as the *soil water matric potential head,* and its absolute value is called the *matric suction head, z* is the distance in the direction of flow.

The more general case of unsteady or transient flow in unsaturated porous media is represented quantitatively by a combination of Darcy's law and the continuity or conservation law for water flow. Richards' equation (Eq. 1.2) combines both of these laws in one formula. Richards' equation of unsaturated flow in its one-dimensional form with the vertical coordinate, z, taken positive upward, is written as

$$\frac{\partial \theta}{\partial t} = \frac{\partial}{\partial z}\left[K(h)\left(\frac{\partial h}{\partial z} + 1\right)\right]$$ [1.2]

A large number of laboratory and field methods have been developed over the years for measuring K as a function of h or θ, which are costly and difficult to implement because of the high spatial and temporal variability of soil properties. In order to circumvent these practical difficulties, developing pedotransfer functions (PTFs) is widely acknowledged (Bouma, 1989; Obalum and Obi, 2013; Teferi et al., 2015a; Tomasella and Hodnett, 2004; Vereecken et al., 2010). PTFs predict the soil hydraulic properties from more easily measured data such as textural information, bulk density and/or few retention points (Pachepsky and Rawls, 2003; Rawls et al., 1982; Wösten et al., 1995).

1.2.2   Remote Sensing of Soil Hydrology

Soil moisture is either measured directly using *in-situ* methods (e.g., gravimetric sampling, time domain reflectrometry (TDR)) (Grayson and Western, 1998; Mohanty et al., 1998) or estimated indirectly through remote sensing (RS) techniques. Due to the high spatial variability over short distances, many point measurements and regionalisation techniques are required to estimate areal values. Recent technological advances in satellite remote sensing have offered the possibility to estimate the aerial moisture content of the top soil by a variety of remote sensing techniques, each with its own strengths and weaknesses. Optical (0.4-3 µm)/Infrared (7-15µm) (Bastiaanssen et al., 2006; Carlson et al., 1994; Gillies and Carlson, 1995) and microwave (1mm - 1m) (Njoku et al., 2003; Ulaby et al., 1996) portions of the electromagnetic spectrum are commonly used to estimate soil moisture.

The estimation of soil moisture using an optical/Infrared approach relies mainly on the use of land surface temperature (Ts), vegetation index (VI), and albedo. A unique relationship was developed by Carlson et al. (1994) and Gillies and Carlson (1995) the estimation of regional patterns of surface soil moisture availability with VI versus Ts using a Soil Vegetation Atmosphere Transfer (SVAT) model. The theory can be referred to as the so-called "universal triangle" concept. A scatter plot of remotely sensed VI versus Ts often results in a triangular shape (Carlson et al., 1994;Price, 1990) or a trapezoid shape (Moran et al., 1994), if a full range of fractional vegetation cover and soil moisture contents is represented in the data (Fig. 2). The emergence of the triangular (or trapezoid) shape in VI/Ts feature space is due to the relative insensitivity of Ts to soil water content variations over areas of dense vegetation, but increased sensitivity over areas of bare soil. The triangle is bounded by a "dry edge" and "wet edge". The dry edge is defined by the location of points of highest temperatures that contain differing amounts of bare soil and vegetation and are assumed to represent conditions of limited surface soil water content and zero evaporative flux from the soil. Likewise, the wet edge represents those pixels at the limit of the maximum surface soil water content.

Microwave sensing encompasses both active and passive forms of remote sensing. A passive microwave sensor detects the naturally emitted microwave energy within its field of view. This emitted energy is related to the temperature and moisture properties of the emitting object or surface. Because the wavelengths are so long, the energy available is quite small compared to optical wavelengths. The fields of view must be large to detect enough energy to record a signal. Thus, most passive microwave sensors are therefore characterized by low spatial resolution (10-100 km). The Advanced Microwave Scanning Radiometer-Earth Observing System (AMSR-E) is one example for passive microwave radiometer. The active microwave sensors receive the backscattering which is reflected from the transmitted microwave which is incident on the ground surface. Synthetic aperture radar (SAR), microwave scatterometers (e.g. ASCAT), radar altimeters etc. are active microwave sensors. They can provide high-resolution data (on the order of tens of meters), but they are rather sensitive to soil surface roughness and vegetation-cover variations. The main advantages of microwave remote sensing over the others are: (1) its ability to penetrate clouds; (2) measurement is less sensitive to land surface roughness or vegetation coverage; (3) measurement is directly related to soil moisture through the soil dielectric constant; and (4) active sensors are independent of solar illumination (Njoku et al., 2003). The sensitivity of dielectric constant to soil moisture is in general decreased by the presence of a vegetation cover due to increased scattering and attenuation of the electromagnetic signal (Tansey and Millington, 2001). However, better soil moisture determination under vegetation cover with L-band microwave was reported by Schmullius and Furrer (1992).

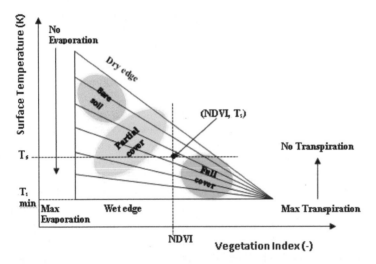

Figure 2 Summary of the main physical properties and interpretations of the remotely sensed estimation of soil moisture based on optical/infrared data using Ts/VI feature space (modified from Lambin (1996) and Sandholt (2002)).

1.2.3 Soil hydrological properties and land use and land cover changes

Land use and management practices play important roles in soil hydrology through the effects of tillage, soil erosion, soil compaction, and geometry and connectivity of pores. Soils under natural forest as compared with soils under human land use, generally exhibit low bulk density, high saturated hydraulic conductivity and high total porosity (Celik, 2005; Teferi et al., 2015b; Teferi et al., 2015; Zhou et al., 2008). These marked differences are resulted from abundant litter cover, organic inputs, root growth and decay, and faunal activity in soils under natural vegetation, which significantly influence soil water retention characteristics and soil structure (Buytaert et al., 2005; Zhou et al., 2008). Intense tillage can result in the breakdown of aggregates, the formation of a plough pan, an increase in soil bulk density, and reductions in soil water storage and transport through pores. In contrast, no tillage or conservation tillage can have the opposite effect, and can improve soil water storage and movement through pores. Thus, efficient soil and water management needs an understanding of soil hydrological properties and land use change effects on these properties.

1.3 PROBLEM STATEMENT

The dominant forms of land degradation that are prevalent in the Upper Blue Nile (UBN) basin include: soil erosion (Hurni and Pimentel, 1993), soil acidification (Teferi et al., 2015), biological soil degradation, plough pan formation caused by the *"maresha"* plough (Biazin et al., 2011; Temesgen et al., 2009), and soil crust formation caused by raindrop impact on soils without vegetation cover. This has led to lower yields and higher costs of fertilizer inputs. Consequently, the food security of present and future generations will be threatened, as the smallholder farmers in the UBN largely depend on the land for their livelihoods.

The major cause of land degradation is human activities (Nyssen et al., 2004). Human activities contributing to land degradation include deforestation, removal of natural vegetation cover, overgrazing and cultivation of steep slopes (Bhan, 1988). Land use and land cover change (LULCC) is the easiest detectable indicator of human influence on the land. LULCC in upland watersheds of Ethiopia influence the livelihoods of the population living in downstream areas by changing critical watershed functions. Such changes also determine, in part, the vulnerability of places and people to climatic, economic, and socio-political pressures. Farmers of the upland watersheds in the Upper Blue Nile basin have already started responding to excessive fertility declines and low productivity per unit of land holdings. They are taking one of the two options: (1) encroach into steep and marginal areas to gain more land to compensate for low yields from their existing holdings (Zeleke and Hurni, 2001), or (2) conversion of their previous croplands in to *Eucalyptus* plantations or change of crop types (e.g. growing crops of poor quality such as engido (*Avena spp.*) in Mt. Choke area). The former involves removal of remnant natural forests from steep lands, which reinforces soil degradation. The total forested area of Ethiopia has decreased substantially during the past half-century, most of it already before 1957. Recent figures show that the country's forest cover has shrunk to less than 4% (TECSULT, 2004).

LULCC has multiple impacts. Beside agriculture and forestry, LULCC affects hydrology (DeFries and Eshleman, 2004; Tekleab et al., 2014; Uhlenbrook, 2007), biodiversity (Hansen et al., 2004a) and ecosystem services (DeFries and Bounoua, 2004). LULCC can also affect climate through its influence on surface energy budgets (e.g. albedo change) and biogeochemical cycling mechanisms (e.g. carbon cycle) (Bounoua et al., 2002; Niyogi et al., 2009; Pongratz et al., 2009). Quantifying LULCC and identifying its multiple impacts has become a topic of paramount importance to the study of global environmental change (Geist and Lambin, 2002). However, there is insufficient awareness among policy makers, planners and the general public to account for the multiple and often interacting determinants and impacts of LULCC in promoting sustainable land management. Therefore, it is important to quantify LULCC and identify its consequences and implications in order to formulate appropriately targeted policy interventions, leading to more sustainable watershed management practices in the headwaters of the upper Blue Nile/Abay basin.

1.4 RESEARCH OBJECTIVES

The general objective of this thesis is to investigate LULCC and its links with soil hydrology, soil degradation and climate variability in the UBN basin. The specific objectives are:

❖ To quantify changes in land use and land cover since 1950s and to identify the determinants of most systematic transitions;

❖ To quantify the spatio-temporal trends in composition and configuration of LULCC;

❖ To quantify wetland dynamics and estimate wetland loss using remote sensing techniques;

❖ To assess effects of land use and land management on selected key soil quality indicators;

❖ To derive and verify pedotransfer functions (PTFs) of soil water retention for high altitude tropical soils in order to assess soil water availability and retention;

❖ To evaluate the impact of LULCC on soil moisture variations using downscaled soil moisture and *in-situ* observations at basin and watershed scales; and

❖ To identify inter-annual and seasonal trends in vegetation conditions at both coarse and fine scales;

❖ To assess the spatial and temporal patterns of NPP and WUE and their climatic controlling factors.

1.5 STRUCTURE OF THE THESIS

Chapter 1 presents the general introduction to the thesis. Chapter 2 presents a brief description the study area. Chapter 3 focuses on changes in land use and land cover since 1950s and the identification of determinants of the most systematic transitions in the Jedeb watershed of UBN basin. Chapter 4 deals with the spatio-temporal trends in composition and configuration of LULCC in the Jedeb watershed. Chapter 5 presents remote sensing-based assessment of wetland dynamics and wetland loss in the Choke Mountain ecosystem. The impacts of LULCC on soil hydrology are presented from Chapter 6 to Chapter 8. Effects of LULCC on soil physical and hydrological properties are described in Chapter 6. In Chapter 7, the impacts of LULCC on soil moisture variation using downscaled soil moisture and *in-situ* observations at basin and watershed scales are presented. Chapter 8 focuses on the derivation and validation of pedotransfer functions (PTFs) of soil water retention for high altitude soils of the Upper Blue Nile region. In Chapter 9, seasonal and inter-annual changes in vegetation condition that are not accounted for in the broad time-scale LULCC analysis are studied. Chapter 10 deals with the climatic impacts and controls of land cover dynamics in the UBN basin. Finally, conclusions and recommendations are presented in Chapter 11.

Chapter 2

STUDY AREA

2.1 BIO-PHYSICAL SETTING

The UBN basin is the Ethiopian part of the Blue Nile also called Abay basin in Ethiopia and covers a total area of 199,812 km^2. It is located in the centre and western part of Ethiopia. It lies approximately between latitude $7°45'$ and $12°46'$ N, and longitude $34°06'$ and $40°00'$ E (Fig. 3 and Fig. 4). The UBN basin is the most important river basin in Ethiopia based on many criteria. First, it accounts for 50% of the total average annual surface runoff while covering only about 17.5% of Ethiopia's land area. Second, it has the country's largest freshwater lake (Lake Tana), covering about 3,000 km^2. Third, the rivers of the UBN basin contribute on average about 62% of the average Nile total discharge at Aswan, Egypt. Finally, it accounts for a major share of the country's irrigation and hydropower potential. The Blue Nile is the largest tributary of the Nile river, the runoff being generated almost all in Ethiopia. Estimates of the mean annual flow range from $45.9 \cdot 10^9$ m^3 a^{-1} to $54 \cdot 10^9$ m^3 a^{-1} (UNESCO, 2004; Conway 2005). Many of the tributaries of the Upper Blue Nile originate from this mountain range. Thus, the Choke Mountain is the water tower of the eastern Nile region serving as headwater of the Upper Blue Nile basin. The annual sediment discharge of the basin at the Sudanese border is between $130 \cdot 10^6$ t a^{-1} and $335 \cdot 10^6$ t a^{-1} (BCEOM, 1998a). In high rainfall areas of the basin, the major source of sediment is rill and gully erosion from sloping cultivated lands resulting from draining excess runoff (Monsieurs et al., 2015; Smit and Tefera, 2011).

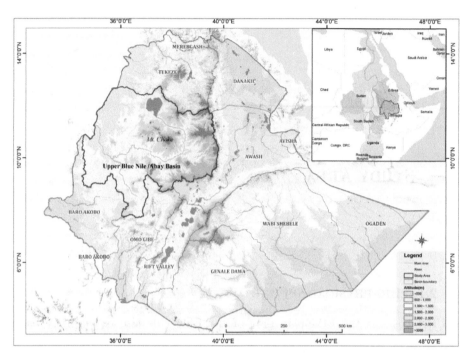

Figure 3 Location map of the Upper Blue Nile (Abay) basin and other river basins of Ethiopia.

The climate of the basin is strongly related to the altitude and the proximity to equatorial monsoon systems. According to Köppen's system, the UBN basin is characterised by tropical, warm temperate and cool highland climate zones. Annual rainfall varies between about 800 mm a^{-1} to 2,220 mm a^{-1}, with a mean of about 1,420 mm a^{-1}. Relatively high rainfall is found in the southern part of the basin and in the centre of the basin and has a shorter, more pronounced wet season. In the eastern and north-western part of the basin, rainfall is relatively low. The mean temperature of the basin is 18.5^{0}C, with mean minimum and maximum daily temperatures of 11.4^{0}C and 25.5^{0}C, respectively. Average annual potential evaporation (Ep) is estimated to be 1,310 mm a^{-1}. The spatial distribution of temperature values is strongly related to altitude. The altitude of the UBN basin ranges from 475 m a.s.l. at the Sudanese border to 4,257 m a.s.l. at the summit of Mt. Guna. The highlands (i.e. altitude greater than 1500 m a.s.l.) and the lowlands are the main landscape units observed in UBN basin. Towards the border with Sudan and extending westwards to the Main Nile, the topography is almost flat or slightly undulating, with just the occasional granite rising above the clay plain.

The geology of the basin is mainly volcanic rocks and Precambrian basement rocks with small areas of sedimentary rock (Conway, 2000). The Highlands of the basin are composed of basic rocks, mainly basalts, while the lowlands are mainly composed of metamorphic rocks and sedimentary rocks of old geological ages. The major soils of the basin are Leptosols, Alisols, Nitisols, Vertisols, Cambisols, and Luvisols, in order of decreasing areal coverage (BCEOM, 1998b). Leptosols (22%) represent the most widely occurring soils within the basin, mostly along the course of the Blue Nile/Abbay River and its main tributaries. They are shallow soils with limited profile development and are usually prone to drought. Alisols (21%) are the second most important soils in terms of area coverage. These soils are reddish brown in colour and have deep profiles (>100 cm). Alisols are mainly derived from basalts, granites and granodiorites and possess favourable drainage, structure and workability. Nitisols (16%) are the third most important soil group within the basin in terms of area. Nitisols are derived from basalts/tuffs and granites/associated felsic materials. These soils are reddish brown in colour, clay to clay loam in texture, well drained and very deep (>200 cm).

Based on the Woody Biomass Inventory and Strategic Planning Project (WBISPP), cultivation occupies 40% of the basin, woodland 24%, shrubland 15%, and grassland 13%. The remaining land cover types occupy about 8% of the basin (TECSULT, 2004). According to Friis et al. (2010) four types of vegetation can be identified in the basin: (1) Combretum-Terminalia broad-leaved deciduous woodland (e.g. *Boswellia papyrifera* and *Anogeissus leiocarpa*), (2) Dry evergreen Afromontane forest (e.g. *Juniperus procera* and *Olea europaea*), (3) Moist evergreen Afromontane forest (e.g. *Euphorbia amliphylla*), and (4) Afroalpine vegetation (e.g. *Lobelia rhynchopetalum* and *Erica arbora*).

Figure 4. Picture of Abay river loaded with sediments and different vegetation types at Tiss Issat falls in Ethiopia (Photo taken in September 2010).

2.2 SOCIO-ECONOMIC SETTING

The administrative structure of Ethiopia is hierarchical, from regional states, to zones, Weredas and Kebeles. According to the current regional structure, the basin is shared between three regional states: Amhara, Oromiya, and Benishangul-Gumuz. The population of the Abay basin is projected to be 22.4 million in July 2015, about 25% of Ethiopia's total population (i.e. 90 million), with an annual growth rate of 2.3% (CSA, 2013). Population density is relatively high (> 150 persons per km^2) in the highlands such as Gojam, Agew Awi and South Gondar and relatively low (< 25 persons per km^2) in the lowlands such as Metekel, Kamashi and Assosa areas (CSA, 2008). The natural environment of the highlands is under great pressure due to increasing demand for fuelwood resulting from the growing population.

The Ethiopian economy is dominated by the agriculture and service sectors each contribute more than 40% to Gross Domestic Product (GDP). About 80% of employment is still concentrated in agriculture and 70% of the raw material requirements of local industries are supplied by the agriculture sector. Thus, the national economy is highly correlated with the performance of the agricultural sector, which suffers from frequent drought, poor cultivation practices and soil degradation. The Abay basin is also predominantly rural in character and the farmers are engaged in small scale and subsistence mixed agriculture. Thus, land and livestock are the basic sources of livelihood. Competition between subsistence agriculture, grazing and fuelwood has led to continuing encroachment onto marginal lands and threatens the sustainability of land resources such as soil, vegetation and water in the basin.

Three main farming systems can be identified in the Abay Basin: (i) the seed-farming complex; (ii) the perennial farming complex; and (iii) shifting cultivation. All three systems are rainfed and involve integration of livestock in varying ways and to varying degrees. Shifting cultivation is recognised in the western part of the basin, along the Sudan border and in the lowland valleys of the Beles, Dura, Abay and Didesa rivers. Sorghum is the main food crop and cotton a cash crop.

Land is a public property in Ethiopia and shall not be subject to sale or other means of exchange. It has been administered by the government since the 1975 radical land reform that has abolished tenant - landlord relationships in Ethiopia. Following the reform, land redistribution was frequent (every two years) and led to land sub-division and fragmentation. A further land distribution is currently being implemented by the Amhara Government with a view to ensuring equitable holdings. The land tenure insecurity is suspected of being a major constraint on the willingness of peasants to adopt long-term soil conservation measures, or to plant perennial crops and trees (Gebremedhin and Swinton, 2003).

PART 1:

MONITORING AND MODELING OF CHANGES IN LAND USE AND

LAND COVER

Figure 5 Progressive expansion of cultivated land on steep slopes of Mt. Choke

Chapter 3

UNDERSTANDING RECENT LAND USE AND LAND COVER DYNAMICS: SPATIALLY EXPLICIT STATISTICAL MODELLING OF SYSTEMATIC TRANSITIONS[1]

3.1 ABSTRACT

The objective of this paper was to quantify long-term land use and land cover changes (LULCC) and to identify the spatial determinants of locations of most systematic transitions for the period 1957-2009 in the Jedeb watershed, Upper Blue Nile Basin. Black and white aerial photographs of 1957 and Landsat imageries of 1972 (MSS), 1986 (TM), 1994 (TM) and 2009 (TM) were used to derive ten land use and land cover classes by integrated use of Remote Sensing (RS) and Geographic Information System (Mengistu et al.). Post-classification change detection analysis based on enhanced transition matrix was applied to detect the changes and identify systematic transitions. The results showed that 46% of the study area experienced a transition over the past 52 years, out of which 20% was due to a net change while

[1]*This chapter is based on* Teferi, E., Bewket, W., Uhlenbrook, S., Wenninger, J., (2013). Understanding recent land use and land cover dynamics in the source region of the Upper Blue Nile, Ethiopia: Spatially explicit statistical modeling of systematic transitions. *Agriculture, Ecosystems & Environment* 165, 98-117.

.

26% was attributable to swap change (i.e. simultaneous gain and loss of a given category during a certain period). The most systematic transitions are conversion of grassland to cultivated land (14.8%) followed by the degradation of natural woody vegetation and marshland to grassland (3.9%). Spatially explicit logistic regression modelling revealed that the location of these systematic transitions can be explained by a combination of accessibility, biophysical and demographic factors. The modelling approach allowed improved understanding of the processes of LULCC and for identifying explanatory factors for further in-depth analysis as well as for practical interventions for watershed planning and management.

3.2 INTRODUCTION

Land use and land cover change (LULCC) has been a key research priority with multi-directional impacts on both human and natural systems (Turner et al., 2007) yet also a challenging research theme in the field of land change science. LULCC can affect biodiversity (Hansen et al., 2004a), hydrology (DeFries and Eshleman, 2004; Uhlenbrook, 2007), and ecosystem services (DeFries and Bounoua, 2004). LULCC can also affect climate through its influence on surface energy budgets (e.g. albedo change) and biogeochemical cycling mechanisms (e.g. carbon cycle) (Bounoua et al., 2002; Niyogi et al., 2009; Pongratz et al., 2009). Hence, it has increasingly become a topic of paramount importance for national and international research programs examining global environmental change (GLP, 2005; Turner et al., 1995).

Many studies highlighted that LULCC is a widespread phenomenon in the highlands of Ethiopia (Amsalu et al., 2007; Bewket, 2002; Teferi et al., 2010; Tegene, 2002; Tsegaye et al., 2010; Zeleke and Hurni, 2001). These studies found different types and rates of LULCC in different parts of the country over the different time periods. In most cases, however, expansion of subsistence crop production into ecologically marginal areas, deforestation and afforestation have been the common forms of transitions. These conversions have apparently contributed to the existing high rate of soil erosion and land degradation in the highlands of Ethiopia (Bewket and Teferi, 2009), which is evident from the numerous gullies in cultivated and grazing lands.

In addition to the few published researches, as mentioned above, there are also some non-published works on LULCC covering different parts of the country. For instance, the Land use Planning and Regulatory Department (LUPRD) of Ministry of Agriculture prepared a generalized LULC map of the 1970s at 1:1,000,000 scale for the whole of Ethiopia through visual analysis of Landsat Multispectral Scanner (MSS) (FAO, 1984). The Abay River Basin Master Plan Project prepared the LULC map of Abay River basin at scales of 1:2,000,000 and 1:250,000 based on Landsat Thematic Mapper (TM) images from 1986 to 1990 (BCEOM, 1998c). Woody Biomass Inventory and Strategic Planning Project (WBISPP) carried out an unsupervised classification of Landsat TM data from 1985 to 1991 to produce LULC map at

1:250,000 scale (TECSULT, 2004). The above mentioned LULC datasets provide a generalized LULC situation for a very large area and are of limited use for a small area such as Jedeb watershed (the site of this study), the scale at which detailed land use plans can be designed. In addition to differences in the levels of detail of these data as determined by their purposes, differences also exist in the approaches and extent of field observations. Therefore, there remains a need for development of a reliable and up-to-date LULC datasets using improved image processing techniques for use in the preparation of detailed land use plans at appropriate spatial scales.

To understand how LULCC affects and interacts with earth systems (e.g. hydrosphere, biosphere, and atmosphere), accurate information is needed on what types of change occur, where and when they occur, and the rates at which changes occur (Lambin, 1997). However, a mere LULCC study does not necessarily lead to improved understanding of the processes of land use change which researchers and policy makers would seek in order to establish effective conservation and management strategies for land resources. Understanding the fundamental processes of land transitions requires detection of dominant systematic land cover transitions (Pontius, 2004; Braimoh, 2006) and spatially explicit land use modelling (Veldkamp and Lambin, 2001; Serneels and Lambin, 2001).

Transitions in land use and land cover can be caused by negative socio-ecological feedbacks that arise from a severe degradation in ecosystem services or from socio-economic changes and innovations (Lambin and Meyfroidt, 2010). Furthermore, transitions can be random or systematic (Braimoh, 2006; Pontius, 2004). Conventionally, random transition refers to episodic processes of change, characterized by abrupt changes (Lambin et al., 2003; Tucker et al., 1991), whereas, systematic transition refers to more stable processes of change that evolve steadily or gradually in response to more permanent forces such as growing population pressure and market expansion (Lambin et al., 2003). From a statistical point of view, random transitions occur when a land cover replaces other land cover types in proportion to the size of other categories. A non-random gain or loss implies a systematic process of change in which land cover systematically targets other land covers for replacement (Pontius, 2004; Alo and Pontius Jr, 2008).

Among the various LULCC modelling approaches, the binary logistic regression models have been used in order to identify the proximate causes and determinants of the most systematic transitions of LULCC (Müller et al., 2011; Schneider and Pontius Jr, 2001; Serneels and Lambin, 2001; Verburg et al., 2004). The major advantages of binary logistic regression models are: (i) they allow testing the presence of links between LULCC and candidate explanatory variables, and (ii) they can predict the location of LULCC, if they are combined with spatially explicit data. Logistic regressions have been used to acquire an improved understanding of determinants of LULCC by previous studies (Schneider and Pontius Jr, 2001; Serneels and Lambin, 2001; Verburg et al., 2004; Müller et al., 2011). In order to explain the spatial patterns of LULCC, the analysis was framed in the context of the Boserupian,

Malthusian, Ricardian, and von Thunen's theories. The Boserupinan (Boserup, 1965) and Malthusian (Malthus, 1826) perspectives relate land use changes to population growth. The Ricardian land rent theory relates land use change to intrinsic land quality (e.g. soil quality, slope, elevation). The von Thunen's perspective links land use change mainly with location-specific characteristics such as cost of access to market centres (Chomitz and Gray, 1996).

In the Upper Blue Nile Basin, notwithstanding past research efforts on LULCC, there remains a knowledge gap in distinguishing systematic transitions and understanding the determinants of these transitions. This study aims at: (i) quantifying changes in land use/ cover over the past 52 years, (ii) identifying the most systematic transitions, and (iii) identifying and quantifying the determinants of most systematic transitions in the rapidly changing environment of the Jedeb watershed in the Blue Nile Basin. The results of this study enable researchers and planners to focus on the most important signals of systematic landscape transitions and allow understanding of the proximate causes of changes. In other words, it is useful for investigating the possible drivers of transitions, and hence to propose site-specific and targeted preventative measures to avoid undesirable impacts of land cover changes (Pontius, 2004; Braimoh, 2006).

3.3 STUDY AREA

The study area, Jedeb watershed, lies between 10°23' to 10°40' N and 37°33' to 37°60' E. It covers an area of 296.6 km² and thus is meso-scale in size. It is situated in the southwestern part of Mount Choke, which is a headwater of the Upper Blue Nile in Ethiopia (Fig. 6). Administratively, the area belongs to two *weredas* (districts; *Machakel* and *Sinan*) of East *Gojam* Zone in *Amhara* Regional State of Ethiopia. The watershed is characterized by diverse topographic conditions with its elevation ranging from 2172 m to nearly 4001 m (Fig. 6), and slopes ranging from nearly flat ($< 2°$) to very steep ($> 45°$). Almost all of the farmers in the Jedeb watershed are engaged in small scale and subsistence mixed agriculture. Thus, land and livestock are the basic sources of livelihood. The livestock population density is higher in upper part of the watershed (*Sinan wereda*) than in the lower part (*Machakel wereda*) as shown in Table 11. The major source of water for both the human and livestock populations is the Jedeb river, with a mean annual stream flow of 716.6 mm a⁻¹ (over the period 1973-2001). The mean annual rainfall varies between 1400 mm a⁻¹ and 1600 mm a⁻¹ based on data from 3 nearby stations (Debre Markos, Anjeni, and Rob Gebeya).

The headwater area as a whole, where the Jedeb river is one of the many tributaries, is a crucial water source area for the Upper Blue Nile. This important water supply zone, as some research results show, is presently one of the most soil erosion-affected parts of the Upper Blue Nile River Basin (Hurni et al., 2005; Bewket and Teferi,

19

2009). Therefore, the region has currently become a focal point of public concerns and received considerable research attention (Tekleab et al., 2011; Uhlenbrook et al., 2010).

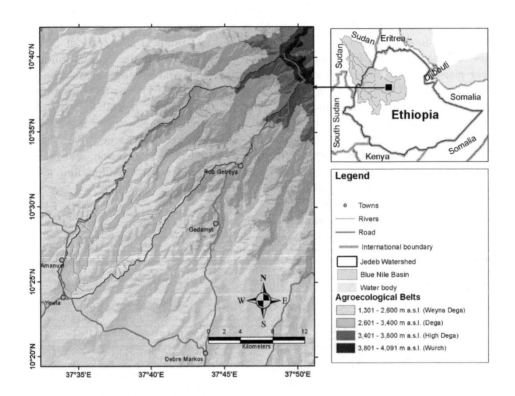

Figure 6 Location map of the study area and topography

3.4 METHODOLOGY

3.4.1 Data used and image pre-processing

Aerial photographs and satellite images: The earliest known sources of land use and land cover data for the study area are black and white aerial photographs taken in 1957. Accordingly, 13 contact prints of aerial photos (Table 1) at a scale of 1:50,000 were obtained from the Ethiopian Mapping Agency (EMA) for the land use and land cover classification purpose. For ground truthing, 15 aerial photographs of the year 1982 at a scale of 1:40,000 were purchased and used (Table 1). For the recent years, Landsat images were used, which were selected based on: (i) availability, (ii) cloud cover percentage, and (iii) correspondence with years of major events in the study area (Table 1). All Landsat images were accessed free of charge from the US Geological Survey (USGS) Centre for Earth Resources Observation and Science (EROS) via http://glovis.usgs.gov/. To assist the land cover classification, a high resolution data set of SPOT-5 2007 was also purchased from the EMA.

Table 1 Description of used aerial photographs and satellite images

	1972	1986[a]	1994	2009[b]
Satellite (sensor)	Landsat MSS	Landsat TM	Landsat TM	Landsat TM
Path/Row	182/52&53	169/53	169/53	169/53
Acquisition date	09/12/1972	28/11/1986	02/11/1994	27/11/2009
Pixel spacing (m)	60	30	30	30
Sun Elevation	44.68	44.96	49.26	50.38
Sun Azimuth	138.94	134.35	127.01	142.25

Aerial photographs of 1957					Aerial photographs of 1982				
26/11/57	18/01/58	24/12/57	16/12/57	15/12/57	01/12/82	01/20/82	01/15/82	01/22/82	01/27/82
		9573	6949	6713			0209	0316	0407
						0144	0208	0315	0406
		9572	6950	6712		0143	0207	0314	
	11630	9571	6951	6711	0069	0142	0206		
2383	11629	9571			0068	0141			

[a] Aerial photographs of 1982 and four topographic maps at a scale of 1:50 000 dated 1984 were used for ground truth data collection to classify the 1986 image.
[b] Since 4% of the 2009 image in the study area was covered by clouds. It was masked out and latter replaced by classification of ASTER (2007) AST_L1B_00303192007080348_17681and AST_L1B_00301302007080346_16466.

Ancillary data: Ancillary data were used to improve accuracy of the classification and to support the interpretation of land cover change. Digital Elevation Terrain Data (DTED-30m) and its derived data sets such as slope supplemented the satellite data in mapping land cover since vegetation classes are often limited to specific agro-climatic conditions. Four topographic maps of the study area at a scale of 1:50,000 dated 1984 were purchased from EMA for the purpose of ground truth data collection. Population data (1984-2007) and vector overlays such as roads, and rivers were obtained from the Central Statistical Agency of Ethiopia (CSA). No national census of population and housing took place prior to 1984 in Ethiopia. Even after 1984 the censuses have been carried out based on administrative boundaries. Hence, remote sensing techniques were used to estimate the population in the watershed, by counting the number of huts from aerial photographs (high resolution satellite image) and by multiplying it by mean occupants per hut derived from available secondary sources. The average family size for 1957 was 4.7 (USBR, 1964); 4.5 average family size for 1986 (Yoseph and Tadesse, 1984) and 4.5 average family size for 2007 (based on sample *kebelles*) (CSA, 2008) were used.

Aerial photograph pre-processing and interpretation: A total of 28 aerial photographs were scanned in A3 format to include the fiducial marks (which were used to establish the interior orientation) with a reasonable geometric (600 dpi) and radiometric resolution (8 bit, grey scale, uncompressed). All photos of a block were scanned in the orientation in which they form the block. To remove the distortion within aerial photographs, caused by terrain relief and the camera, the aerial photos were orthorectified using Digital Terrain Elevation Data (DTED-30m), fiducial marks (mm), the camera's focal length (mm) and Ground Control Points (GCPs) as input in ERDAS IMAGINE software. Since there was no access to the camera calibration report, the fiducial coordinates were measured from the aerial photographs such that the origin of the coordinate system is the principal point of the aerial phtotograph. On each aerial photograph eight to twelve well-distributed X and Y coordinates of the GCPs were collected from digital orthorectified SPOT image (5m) and Z coordinates were taken from (DTED-30m). Rural churches were used as GCPs. The root mean-square error (RMS) of GCP locations was 0.4 pixels. The accuracy of the orthophotos was further assessed by overlaying the orthorectified photos over the available georefernced topographic map. The multiple orthophotos were then mosaiced to form a seamless image. The complete procedure is displayed in Figure 7.

Visual interpretation of various features on the mosaiced orthophotos was done based on shade, shape, size, texture and association of features to identify the features. Unrectified stereo-pairs of airphotos were used when a detailed 3D view was required to identify the features. After subsequent screen-digitizing of the user-defined land-cover units, GIS topology was employed to manage coincident geometry and ensure data quality (Fig. 8a & b).

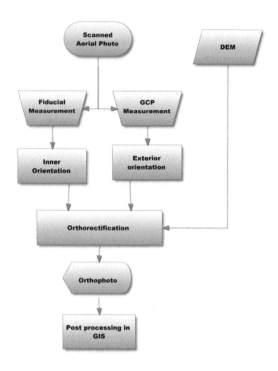

Figure 7 Flowchart of the 1957 aerial photograph processing

Satellite Image Pre-processing, classification and accuracy assessment: All the scenes obtained from the EROS Data Centre were already georeferenced to the Universal Transverse Mercator (UTM) map projection (Zone 37), WGS 84 datum and ellipsoid. However, in some parts of the study area, there were significant discrepancies between these imageries, and the underlying GIS base layers which were extracted from high resolution SPOT-5 imagery. The nonaligned scenes were then georectified to the underlying base layers by using control points. Re-projection to the local projection system was made (UTM, map projection; Clarke 1880, Spheroid; and Adindan Datum) (Fig. 8c &d).

Since the Jedeb watershed is highly rugged and mountainous, all Landsat images were topographically corrected using C-correction (Teillet et al., 1982) to reduce effects that result from the differences in solar illumination (McDonald et al., 2000; Hale and Rock, 2003). A freely available IDL extension was used for the C-correction (http://mcanty.homepage.t-online.de/software.html). Atmospheric and radiometric correction of all Landsat images acquired for the different times were also made using a fully image-based technique developed by Chavez (1996) known as the COST model; this model derives its input parameters from the image itself. In a next step, temporal normalization of all the images was achieved by applying regression equations to the 1972, 1986, 1994, and 1998 imageries which predict what a given

23

Brightness Value (BV) would be if it had been acquired under the same conditions as the 2009/11/29 TM reference scene. This pre-processing step reduces pixel brightness value variations caused by non-surface factors and enhances variations in pixel BV between dates related to actual changes in surface conditions. The 2009/11/29 TM image was selected as the reference scene to which the other images were normalized, because it was the only year for which quality *in situ* ground reference data were available.

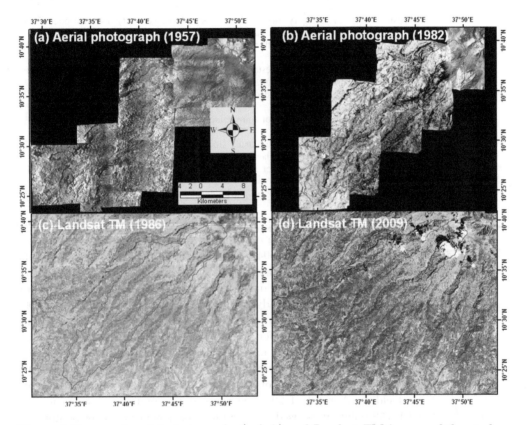

Figure 8 Processed aerial photographs (a & b) and Landsat TM images of the study area with 742 band combination (c & d)

A hierarchical land use classification was derived from the authors' prior knowledge of the study area and based on the most popular scheme of the U.S. Geological Survey Land Use/Cover System devised by Anderson et al. (1976). Table 2 and Figure 9 contain a list of all land cover types present in the study area that could be clearly identified from the satellite images and aerial photographs. All classes have an unambiguous definition so that they are mutually exclusive.

Table 2 Description of land use and land cover classes

No	Land use and land cover classes	Description	Code
1	Grassland	Landscapes that have a ground story in which grasses are the dominant vegetation forms.	GL
2	Afroalpine grassland	High altitude (>3500 m asl) herbaceous vegetation typically consisting of *Lobeliaceae* (Hedberg, 1951).	AGL
3	Shrubs and bushes	This category includes low woody plants, generally less than three meters in height, usually with multiple stems, growing vertically.	SHB
4	Cultivated land	Areas covered with annual crops followed by harvest and bare soil period	CL
5	Riverine Forest	Trees and shrubs established along the banks of streams, rivers, and open bodies of water.	RF
6	Woodland	A continuous stand of a single storey trees with a crown density of between 20 - 80%.	WL
7	Plantation Forest	Areas composed of transplanted seedlings of *Eucalyptus globules* and *Cupresus spp.*	PF
8	Marshland	Periodically or continually flooded wetlands characterized by non-woody emergent plants	ML
9	Barren land	Areas with little or no vegetation cover consisting of exposed soil and/or bedrock	BL
10	Ericaceous Forest	*Ericaceae, Hypericum quartiniannm*	EF

A hybrid supervised/ unsupervised classification approach was integrated with successive Geographic Information System operations (spatial analysis) to classify the imageries of 1972 (MSS), 1986 (TM), 1994 (TM) and 2009 (TM). First, to determine the spectral classes Iterative Self-Organizing Data Analysis (ISODATA) clustering was performed. Ground truth (reference data) was collected to associate the spectral classes with the cover types of the already defined classification scheme for the 2009 (TM) imagery from field observations. Among other sources of information, reference data was collected from previous maps for the 1994 image (Zeleke and Hurni, 2001). Reference data for the 1986 image were based on aerial photo interpretation of 1982 as well as topographic maps (1:50,000 scale) of 1984. Reference data for the 1972 image classification were collected from already classified maps and in-depth interview of local elders. A total of 2277 reference data points for each year were obtained, of which 759 points were used for accuracy assessment and 1518 points were used for classification. Training sites were developed from the ground truth data collected to generate a signature for each land cover type. Signature separability was determined and it was found that the spectra of marshland were confused with grassland, urban land covers were confused with barren land and cultivated land during the preliminary classification. To tackle this problem, urban areas were digitized and masked out from the image. Similarly, marshlands were identified and masked out from the original image. The remaining part of the image was then classified using the Maximum Likelihood algorithm to identify grassland, cultivated land, shrubs and bushes, woodland, plantation forest and *Ericaceous* forest (Fig. 10).

Figure 9 Views of land use and land cover types in the Jedeb watershed (a - f)

To assess the accuracy of thematic information derived from 1972 (MSS), 1986 (TM), 1994 (TM) and 2009 (TM) imageries, the *design-based statistical inference* method was employed which provides unbiased map accuracy statistics (Jensen, 2005). The accuracy of the map interpreted from the 1957 aerial photographs was assessed using *confidence-building assessment* which involves the visual examination of the classified map by knowledgeable local people to identify any gross error. Stratified random sampling design was used to collect 759 reference data and to make careful observation during field visit of accessible areas. The number of samples required for each class was adjusted based on the proportion of the class and inherent variability within each category.

Discrete multivariate analytical techniques were used to statistically evaluate the accuracy of the classified maps (Foody, 2002). As there is no single universally accepted measure of accuracy, a variety of indices were calculated (e.g. overall accuracy, producer's accuracy, user's accuracy and kappa analysis) (Stehman, 1997).

3.4.2   Extent, rate and trajectories of change

Gross gains, losses and persistence: Post-classification comparison change detection was made to determine the change in land use/ cover between two independently classified maps from images of two different dates (Jensen, 2005). Although this technique has a few limitations, it is the most common approach to compare maps of different sources, as it provides detailed "from-to" change class information (Coppin et al., 2004; Jensen, 2005), and it does not require data normalization between two dates (Singh, 1989). The traditional cross-tabulation matrix (transition matrix) was computed using overlay functions in ArcGIS 9.3 software and Pivot Table function in Excel to analyse land use/ cover transitions. Analysis of gains, losses, persistence, swap and net change were carried out for each of the periods: 1957-1972, 1972-1986, 1986-1994, 1994-2009 and 1957-2009.

The computed transition matrix consists of rows that display categories at time 1 and columns that display categories at time 2 (Table 5). The notation P_{ij} is the proportion of the land that experiences transition from category i to category j. The diagonal elements (i.e. P_{jj}) indicate the proportion of the landscape that shows persistence of category j. Entries off the diagonal indicate a transition from category i to a different category j. According to Pontius (2004), the proportion of the landscape in category i in time 1 (P_{i+}) is the sum of P_{ij} over all j. Similarly the proportion of the landscape in category j in time 2 (P_{+j}) is the sum of P_{ij} over all i. The losses were calculated as the differences between row totals and persistence. The gains were calculated as the differences between the column totals and persistence. Persistence of land covers were also analysed based on methods outlined in Braimoh (2006).

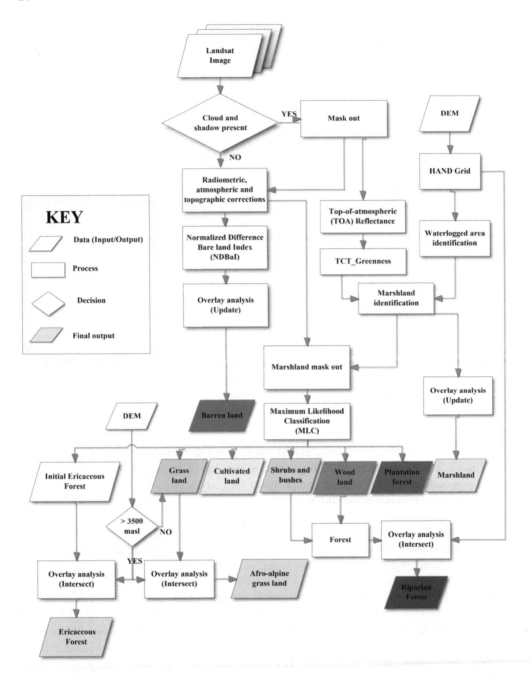

Figure 10 Flow chart showing the steps for identifying land covers types from satellite images

The vulnerability of land cover classes to transition were assessed by calculating the loss-to-persistence ratio denoted by l_p. Higher tendency of land cover transition to other categories than persist is expressed by values of l_p greater than one. Similarly, values of gain-to-persistence ratio denoted by g_p greater than one indicate more gain than persistence.

Net change and swap: The difference between gain and loss is the net change (absolute value), denoted D_j. However, the net change underestimates the total change on the landscape since it fails to capture the "swap". Swap implies simultaneous gain and loss of a category on the landscape. The amount of swap of land class j, denoted S_j, was calculated as two times the minimum of the gain and loss because to create a pair of grid cells that swap each grid cell that gains is paired with a grid cell that loses. The total change for each land class, denoted C_j, was calculated as either the sum of the net change and the swap or the sum of the gains and losses.

$$D_j = |P_{+j} - P_{j+}| \tag{3.1}$$

$$S_j = 2 \times MIN(P_{j+} - P_{jj}, P_{+j} - P_{jj}) \tag{3.2}$$

$$C_j = D_j + S_j = P_{j+} + P_{+j} - 2P_{jj} \tag{3.3}$$

Rate of change: The annual rate of change of land use/ cover at different periods was computed using the formula derived from the Compound Interest Law due to its better estimation and biological meaning (Puyravaud, 2003).

$$r = \left(\frac{1}{t_2 - t_1}\right) * \ln\left(\frac{A_2}{A_1}\right) \tag{3.4}$$

where r is annual rate of change, and A_1 and A_2 are the area coverage of a land cover at time t_1 and t_2, respectively. This equation provides a standard method for making land use and land cover change comparisons that are insensitive to the differing time periods between observation dates.

Trajectories of change: Trajectory analysis was made only for forest, non-forest, and plantation classes because of their ecological importance. Therefore, SHB, RF, EF and WL classes were reclassified as forest (e.g. F) category and GL, AGL, CL, ML and BL classes were reclassified as non-forest category (e.g. N). The number of permutations of three classes, taken five at a time when each can be repeated five times is given by: $NT = NC^d$, where NT is the number of trajectories, NC is the number of land cover classes and d is the number of observation dates. With three land cover classes (forest, non-forest, and plantation) and five temporal image dates (1957, 1972, 1986, 1994 and 2009), 114 out of 243 possible forest land cover change trajectories were found. Finally, similar trajectories were clustered to result in seven classes (Figure 7 and Table 3).

Table 3 Land use and land cover change trajectories between 1957 and 2009

No	Description	Trajectories*
1	Stable forest	FFFFF
2	Afforestation/reforestation	FFFFP,FFFNP,FFFPF,FFFPP,FFNFP,FFNNP,FFNPP,FFPFP, FFPNP,FFPPP,FNFFP,FNFNP,FNFPF,FNFPP,FNNFP,FNNNP,FNNPF,FNNP P,FNPFP,FNPPP,FPPPP,NFFFP,NFFNP,NFFPF,NFFPP,NFNFP,NFNNP,NFN PP,NFPFP,NFPPP,NNFFP,NNFNP,NNFPP,NNNFP,NNNNP,NNNPF,NNPFP, NNPNP,NNPPF,NNPPP,NPPFP,NPPPP
3	Recent deforestation	FFFFN,FFFPN,FFNPN,FFPFN,FFPPN,FNFPN,FNNPN,FNPFN,FNPPN ,FPPPN,NFFFN,NFFPN,NFNFN,NFNPN,NFPFN,NFPPN,NNFFN,NNFPN
4	Old deforestation	FFFNN,FFNNN,FFPNN,FNNNN,FNPNN,NFFNN,NFNNN, NFPNN,NNFNN,NNNPP,NNPNN,NPNNN,NPPNN
5	Regrowth	FFFNF,FFNFF,FFNNF,FFNPF,FFPFF,FFPNF,FFPPF,FNFFF, FNFNF,FNNFF,FNNNF,FNPFF,FNPNF,FNPNP,FNPPF,NFFFF, NFFNF
6	Regrowth with new clearing	FFNFN,FNFFN,FNFNN,FNNFN
7	Non forest	NNNNN

*The sequence represents the time periods 1957, 1972, 1986, 1994, and 2009. "F" stands for forest class, "N" stands for non-forest class, and "P" stands for plantation.

3.4.3 Detecting the most systematic transitions

The traditional way of identifying the most prominent types of transition is by ranking each conversion between classes after summing up the total area changed during each period. However, such a technique fails to consider the presence of the largest categories. Consequently, even the random process of land use/ cover change would cause a large transition of dominant categories. Therefore, interpreting the transitions relative to the sizes of the categories is crucial in order to identify systematic transitions. This was carried out based on methods outlined by Pontius (2004). First expected gains and expected losses (L_{ij}) that would occur if random changes occurred were determined (Eq. 3.5 and Eq. 3.6). Values obtained were then used to compute the difference between the observed and expected transition under a random process of gain $(P_{ij} - G_{ij})$ or loss $(P_{ij} - L_{ij})$, denoted D_{ij} and the ratio, denoted R_{ij}. The values of D_{ij} indicate the tendency of category "j" to gain from category "i" (focus on gains) and the tendency of category "i" to lose to the category "j" (focus on losses). Large positive or negative deviations from zero indicate that systematic inter-category transitions, rather than random transitions occurred between two land classes. The magnitude of R_{ij} indicates the strength of the systematic transition. A large ratio means that the transition is systematic.

$$G_{ij} = (P_{+j} - P_{jj})\left(\frac{P_{i+}}{100 - P_{j+}}\right), \quad \forall\, i \neq j \tag{3.5}$$

$$L_{ij} = (P_{i+} - P_{ii})\left(\frac{P_{+j}}{100 - P_{+i}}\right), \quad \forall\, i \neq j \tag{3.6}$$

3.4.4 Logistic regression modelling of the most systematic transitions

Logistic regression was used to identify and quantify the relations between the locations of most systematic transitions and a set of explanatory variables. Two processes of systematic land use/ cover transitions have been identified: (i) forest-to-grassland conversion, (ii) grassland-to-cultivated land. *Ericaceous* forest-to-grassland, shrubs and bushes-to-grassland, and riverine forest-to-grassland transitions were aggregated to form a forest-to-grassland dependent variable, as these transitions have been identified as most systematic. The dependent variable is a binary presence (specific land use conversion) or absence event (all other) for the period 1957–2009. The logistic function gives the probability of observing a given land use conversion as a function of the independent variables in Table 9. If p is a *probability* then $p/ (1 - p)$ is the corresponding *odds*, and the logit of the probability is the logarithm of the odds. The linear logistic model has the form:

$$logit\,(p) = \log\left(\frac{p}{1-p}\right) = \beta_0 + \beta_1 X_1 + \beta_2 X_2 + ... + \beta_n X_n \tag{3.7}$$

where β_0 is the intercept and β_n are the logit coefficients.
The probability can be expressed in terms of independent variables as

$$p = \frac{\exp(\beta_0 + \beta_1 X_1 + \beta_2 X_2 + ... + \beta_n X_n)}{1 + \exp(\beta_0 + \beta_1 X_1 + \beta_2 X_2 + ... + \beta_n X_n)} \tag{3.8}$$

In order to assess the relative importance of the explanatory variables in determining most systematic conversions, the independent variables were standardized to zero mean and unit standard deviation, using the equation

$$Z = (x - \mu)/\sigma \tag{3.9}$$

where Z is the standardized variable, x is the original value of the independent variable, μ is the mean and σ is the standard deviation.

The goodness of fit of the logistic regression models was evaluated using the relative operating characteristics (ROC) (Pontius Jr and Schneider, 2001). The ROC value above 0.5 is statistically better than random, while a ROC value greater than 0.7 is acceptable for land use change modelling (Lesschen et al., 2005).

Generation of spatially explicit independent variables: The variables *elevation, slope, aspect and Topographic Wetness Index (TWI)* were derived from Digital Elevation Terrain Data (DTED-30m). Since the initially derived aspect variable was based on circular categorical values, it was again converted into a northness [cos(aspect)] and eastness [sin(aspect)] for the model input. In sin(aspect) map values close to 1 indicate east-facing slopes, while in cos(aspect) map values close to 1 represent north-facing slopes. Accessibility was calculated in terms of travel time to targets (roads and towns/markets). Anisotropic *Travel time to roads* and anisotropic *travel time to towns/markets* were calculated based on Tobler's hiking function (Tobler, 1993), as the direction of slopes affects the efforts needed to cross an area especially in mountain environments. Euclidean *Distance from forest edges* was calculated using *centre versus neighbour* measure of pattern on a 3x3 pixel window. The likelihood for agricultural use is dependent on proximity to rivers. Therefore, Euclidean distance of locations to permanent rivers was calculated. In order to account for the impact of the population pressure in determining the land use change, initial population density or population potential and change in population density or population potential between 1957 and 2007 were calculated. Population potential surfaces were generated based on population data collected at village (*Kebelle*) level for 2007 and 1957 (back-projected) using inverse distance weighting. However, population potential variables were found to be irrelevant variables because they had no explanatory power. In contrast, population density surfaces generated from aerial photos and satellite imagery were able to explain the logistic regression models. Hence, the population potential variables were removed from the analysis.

Prior to logistic regression modelling, violations of three major assumptions were tested: linearity of the logit, spatial dependence, and multi-collinearity. In order to minimize the effect of spatial auto-correlation, an initial random sampling of 6595 observations (i.e. 4% out of 329,729 observations) was carried out for each model. Subsequently, 1876 observations for grassland-to-cultivated land conversion and 764 observations for forest-to-grassland conversion model were sampled with an equal number of 0 and 1 observations of the dependent variable. Since logistic regression assumes a linear relationship between the logit of the independent and the dependent variable, linearity of the logit was tested using Box-Tidwell transformation (Menard, 2009). Variables with Spearman rank correlation coefficients of 0.8 or higher were removed as recommended by Menard (2001) to remove multi-collinearity effect. Besides, outliers were detected and removed by observing very large Cook's distance scores (Cook, 1977) from a scatter plot of Cook's distance scores versus sample codes.

Table 4 List of variables included in the binary logistic regression

Independent variables	Abbreviation	Units	Proxy for
Topography related variables			
Slope	SLP	degrees	Diffuse solar radiation
Elevation	ELEV	m	Mean annual air temperature
Sine of aspect	SIN_ASP	-1 to 1	Eastness
Cosine of aspect	COS_ASP	-1 to 1	Northness
Topographic wetness index	TWI	Index	Moisture accumulation
Distance variables			
Distance to forest edge	D_FOREDGE57	m	Likelihood for forest loss
Distance to river	D_RIV	m	Likelihood for agriculture use
Travel time to roads	TT_ROAD	min	Accessibility*
Travel time to towns (market)	TT_TOWN	min	Accessibility*
Demographic variables			
Population density in 1957	PD_57	Inha./km^2	Likelihood of forest loss
Change in population density (2007-1957)	PD07_57	Inha./km^2	Likelihood of forest loss

*Measure of accessibility is defined as the ability for contact with sites of economic or social opportunities (Deichmann, 1997)

3.5   RESULTS AND DISCUSSION

3.5.1   Accuracy Assessment

Figure 5 depicts the classified maps for 1957, 1972, 1986, 1994, and 2009. According to the confusion matrix report (Table 4) 95.65% overall accuracy and a Kappa Coefficient (*khat*) value of 0.94 were attained for the 2009 classified map. Similarly, overall classification accuracies achieved were 89.25% (with a *khat* of 0.84) for the 1957, 87.67% (with *khat* of 0.81) for the 1972, 91.47% (with *khat* of 0.89) for the 1986, and 94.17% (with *khat* of 0.91) for the 1994 image classifications. Applying the methods of Congalton and Green (2009) the above results represent strong agreement between the ground truth and the classified classes. In general, the maps met the minimum accuracy requirements to be used for the subsequent post-classification operations such as change detection (Anderson et al., 1976).

Table 5 Confusion matrix (error matrix) for the 2009 classification map

Classified data	GL	AGL	SHB	CL	RF	WL	PF	ML	BL	EF	Row total	User's accuracy
GL	190	1	1	1	2	0	0	0	1	0	196	97%
AGL	0	67	0	1	0	0	0	0	0	1	69	97%
SHB	0	0	18	0	1	3	3	0	0	0	25	72%
CL	2	0	0	323	0	0	0	0	1	0	326	99%
RF	0	0	0	0	13	0	1	1	0	0	15	87%
WL	0	0	0	0	1	16	2	0	0	0	19	84%
PF	0	0	0	0	0	1	45	0	0	0	46	98%
ML	1	1	2	0	0	0	0	34	0	0	38	89%
BL	2	2	0	1	0	0	0	0	17	0	22	77%
EF	0	0	0	0	0	0	0	0	0	3	3	100%
Column total	195	71	21	326	17	20	51	35	19	4	759	
Producer's accuracy	97%	94%	86%	99%	76%	80%	88%	97%	89%	75%		

Overall accuracy = 95.65%, Khat = 94 %

3.5.2 Extent, rate and trajectories of land use and land cover change

Gross gains, gross losses and persistence

Table 5 depicts the proportion of each land use/ cover class that made a transition from one category to another for each of the study periods. During the whole period 1957-2009, cultivated land and grassland were the major proportions as compared to the other land use/ cover classes. Both the highest gain (21.9%) and the highest loss (13.5%) in cultivated land occurred during the period 1957-1972. The highest gain in grassland (14.1%) was experienced during the period 1957-1972, while the highest loss (7.2%) was experienced during the period 1994-2009. In the period 1957-1972, a decrease in the following natural woody vegetation covers was observed: shrubs and bushes (declined by 53.7%); riverine forest (declined by 47.5%); wood land (declined by 27.5%); and *Ericaceous* forest (declined by 22.2%) from the initial states. Although a decline in natural woody vegetation was observed in all periods, the highest declines in percentage changes of shrubs and bushes, riverine forest and woodland were observed during the period 1957-1972. In contrast, cultivated land showed a relative growth of 15.7% from the original state. During the period 1972-1986, cultivated land experienced more gain (17.7%), whereas grassland experienced more loss (17.4%). Cultivated land, plantation forest, and barren land showed relative growths of 7.6%, 600% and 80%, respectively, whilst the other categories showed relative declines from their initial states; i.e. 1972 (Fig. 11 and Fig. 12). During the period 1986-1994, plantation forest and cultivated land showed relative increase by 342.9% and 4.2%, respectively, while the remaining categories showed declines. During the periods of 1972-1986 and 1986-1994 *Eucalyptus globules* plantation expanded significantly. The decrease in percentage change of barren land only in the period 1986-1994 indicates that the environment was recovering from the devastating drought of 1984/1985. In the period 1994-2009, the percentage of bare land, plantation forest and grassland showed increase by 140%, 9.7% and 4.9%, respectively, while riverine forest declined by 38.9%, shrubs & bushes declined by 35%, marshland declined by 8%, afro-alpine grassland declined by 7.7%, and wood land declined by 7.1%. The increase in percentage change of cultivated land has stopped since 1994. This is likely because of lack of suitable lands for further expansion, or perhaps there are more strict local government rules and regulations with respect to land tenure and land use rights, prohibiting expansion of cultivation at one's own will. This finding is in line with Tegene (2002) who concluded that there was no significant expansion of cropland in mountainous catchments (i.e. Derekolli) in Northern Ethiopia.

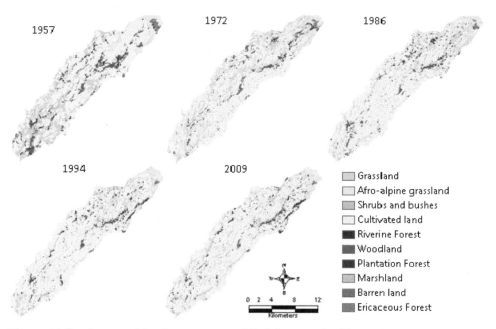

Figure 11 Land use and land cover map of Jedeb watershed in 1957, 1972, 1986, 1994 and 2009

The diagonal entries in Table 5 show the percentage of landscape that remained unchanged. About 56%, 62%, 71%, and 75% of the landscape persisted or 44%, 38%, 29% and 25% of the landscape has changed during the period 1957-1972, 1972-1986, 1986-1994, and 1994-2009, respectively, indicating that persistence dominates in all periods. Cultivated land, grassland, afro-alpine grassland together accounted for 50%, 58%, 67%, and 71% of the persistence of the landscape in the respective periods. Both the loss-to-persistence ratio (l_p) and gain-to-persistence ratio (g_p) for grassland, wood land, and marshland are greater than 1 during the period 1957-2009 suggesting that these classes have a tendency to lose or gain rather than to persist. Figure 12a depicts this pattern for grassland. The loss-to-persistence ratio is greater than 1 for riverine forest and shrubs and bushes, indicating that they experienced a higher tendency to lose than persist in the period 1957-2009. Cultivated land showed a tendency to persist or gain rather than lose (Figure 13b) suggesting more intensive as well as extensive cultivation. Moreover, the persistence of cultivated land is widespread in areas where population density is highest. Afro-alpine grassland tends to persist rather than lose or gain; riverine forest and shrubs and bushes tend to lose rather than gain/persist; and grassland, woodland and marshland tend to rather gain or lose than persist.

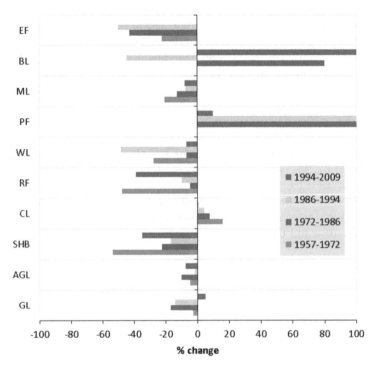

Figure 12 Percentage of change of each category; see Table 2 for the definition of the abbreviations

Net change and swap
Swap land change dynamics accounted for 81%, 85%, 82%, and 92% of total landscape change for periods 1957-1972, 1972-1986, 1986-1994, and 1994-2009, respectively. During the whole period 1957-2009, changes in plantation forest and *Ericaceous* forest consist of only a net change, whereas changes in all other categories consist of both net and swap type of changes. In the Jedeb watershed as a whole during this period the change attributable to location (swap = 25.9%) is larger than the change attributable to quantity (net change = 20.1%). During the period 1957-1972, changes in shrubs and bushes, cultivated land, riverine forest, wood land, marshland and *Ericaceous* forest consisted of both net and swap type changes,

whereas changes in grassland, afro-alpine grassland and bare land consisted mainly of swap type of change. For example, the net change of grassland (0.6%) perhaps indicates as if there was a lack of change in the landscape because it fails to capture the swapping component of change (28.2%). The change attributable to the net change is highest for marshland (58% of total change for marshland); whereas the change attributable to swap is highest for grassland (98% of the total change for grassland). During the period 1972-1986, all categories consisted of both net and swap types of change with the exceptions for riverine forest and wood land which consisted of mainly swap type of change. In the period 1986-1994, the change in *Ericaceous* forest consisted of almost pure net change, while the change in the rest categories, with the exceptions for afro-alpine grassland and marshland, consisted of both swap and net changes. During the period 1994-2009, changes in grassland, cultivated land, wood land, plantation forest, and marshland experienced swapping change dynamics, whereas the changes in all other categories consisted of both swap and net changes.

Rate of land use and land cover change and trajectories of forest cover

The rate of land use and land cover change throughout the 52 years studied showed periodic fluctuations (Table 6). The annual rate of increase of cultivated land slowed down from 1.2% a^{-1} in 1957 to 0.1% a^{-1} in 2009, suggesting a decrease in agricultural extensification over time. Higher annual rates of decline of grassland were observed during the periods 1972-1986 (-1.7% a^{-1}) and 1986-1994 (-1.9% a^{-1}). The succession of droughts during those periods could be the major factor for those changes.

Table 6 Land use and land cover change flow matrices for each period (percentage)

1957/1972	GL	AGL	SHB	CL	RF	WL	PF	ML	BL	EF	P_{i+}	Loss	C_i	S_i	D_i	l_p	g_p
GL	5.8	0.0	0.3	13.7	0.2	0.2	0.0	0.1	0.1	0.0	20.5	14.7	28.8	28.2	0.6	2.5	2.4
AGL	0.2	4.2	0.3	1.0	0.2	0.1	0.0	0.0	0.0	0.2	6.2	2.0	3.7	3.4	0.3	0.5	0.4
SHB	2.3	0.0	0.6	3.3	0.0	0.4	0.0	0.0	0.0	0.0	6.7	6.1	8.6	5.0	3.6	10.2	4.2
CL	8.9	1.1	1.4	39.9	0.7	0.9	0.0	0.1	0.3	0.0	53.4	13.5	35.4	27.0	8.4	0.3	0.5
RF	1.2	0.1	0.0	1.6	1.0	0.1	0.0	0.0	0.0	0.0	4.0	3.0	4.1	2.2	1.9	3.0	1.1
WL	1.0	0.0	0.4	1.5	0.0	1.0	0.0	0.0	0.0	0.0	3.9	2.9	4.8	3.8	1.0	2.9	1.9
PF	0.0	0.0	0.0	0.0	0.0	0.0	0.0	0.0	0.0	0.0	0.0				0.0		
ML	0.4	0.0	0.1	0.4	0.1	0.0	0.0	2.8	0.0	0.0	3.9	1.1	1.4	0.6	0.8	0.4	0.1
BL	0.1	0.0	0.0	0.4	0.0	0.0	0.0	0.0	0.0	0.0	0.5	0.5	1.0	1.0	0.0		
EF	0.0	0.4	0.0	0.0	0.0	0.0	0.0	0.0	0.0	0.5	0.9	0.4	0.6	0.4	0.2	0.8	0.4
P_{+j}	19.9	5.9	3.1	61.8	2.1	2.9	0.1	3.1	0.5	0.7	100	44.2	44.2	35.8	8.4		
Gain	14.1	1.7	2.5	21.9	1.1	1.9	0.1	0.3	0.5	0.2	44.3						

1972/1986	GL	AGL	SHB	CL	RF	WL	PF	ML	BL	EF	Pi+	Loss	C_i	S_i	D_i	l_p	g_p
GL	5.2	0.0	0.7	12.5	0.4	0.4	0.2	0.4	0.1		19.9	14.7	26.0	22.6	3.4	2.8	2.2
AGL	0.1	4.0	0.2	1.2	0.0	0.1	0.0	0.1	0.0	0.1	5.9	1.9	3.1	2.5	0.6	0.5	0.3
SHB	0.6	0.2	0.6	1.3	0.1	0.3	0.0	0.0	0.0		3.1	2.6	4.4	3.6	0.8	4.5	3.2
CL	8.8	0.5	0.6	48.8	0.4	0.9	0.4	0.8	0.7	0.0	61.8	13.0	30.7	26.0	4.7	0.3	0.4
RF	0.4	0.1	0.0	0.7	0.8	0.0	0.0	0.0	0.0	0.0	2.1	1.3	2.5	2.4	0.1	1.7	1.6
WL	0.5	0.1	0.2	1.0	0.0	0.9	0.0	0.0	0.0	0.0	2.9	2.0	3.8	3.5	0.2	2.2	2.0
PF							0.1				0.1	0.0	0.7	0.0	0.7	11.0	
ML	0.7	0.1	0.1	0.5	0.2	0.1	0.0	1.3	0.0	0.0	3.1	1.8	3.2	2.8	0.4	1.4	1.1
BL	0.1	0.0	0.0	0.3	0.0	0.0	0.0	0.0	0.0		0.5	0.4	1.3	0.9	0.4	15.7	30.6
EF	0.0	0.3	0.0	0.1		0.0		0.0	0.0	0.3	0.7	0.3	0.4	0.2	0.2	1.0	0.3
P_{+j}	16.5	5.3	2.4	66.5	2.0	2.7	0.7	2.7	0.9	0.4	100	38.0	38.0	32.3	5.7		
Gain	11.3	1.3	1.8	17.7	1.2	1.8	0.7	1.4	0.9	0.1	38.0						

1986/1994	GL	AGL	SHB	CL	RF	WL	PF	ML	BL	EF	Pi+	Loss	C_i	S_i	D_i	l_p	g_p
GL	6.5	0.0	0.5	8.1	0.3	0.1	0.4	0.5	0.1		16.5	10.0	17.8	15.4	2.3	1.5	1.2
AGL	0.0	3.8	0.0	1.1	0.0	0.0	0.2	0.1	0.0	0.0	5.2	1.4	2.8	2.7	0.1	0.4	0.4
SHB	0.4	0.1	0.7	0.7	0.0	0.2	0.2	0.0	0.0	0.0	2.4	1.7	3.0	2.6	0.4	2.6	2.0
CL	6.1	0.9	0.6	56.5	0.4	0.3	1.3	0.4	0.1	0.0	66.5	10.0	22.7	19.9	2.8	0.2	0.2
RF	0.2	0.1		0.5	0.9		0.2	0.1	0.0		2.0	1.1	2.0	1.8	0.2	1.2	1.0
WL	0.2	0.0	0.2	0.9	0.0	0.7	0.5	0.0	0.0	0.0	2.6	2.0	2.7	1.4	1.3	3.0	1.1
PF	0.0	0.0	0.0	0.6	0.0	0.0	0.1	0.0	0.0		0.7	0.6	3.6	1.2	2.4	5.7	27.8
ML	0.4	0.0	0.0	0.7	0.1	0.0	0.1	1.4	0.0		2.7	1.3	2.3	2.1	0.2	0.9	0.8
BL	0.4	0.1	0.0	0.2	0.0	0.0	0.0	0.0	0.2		0.9	0.7	1.0	0.6	0.4	3.9	1.6
EF		0.2	0.0	0.0	0.0	0.0	0.0	0.0		0.2	0.4	0.2	0.2	0.0	0.2	1.1	0.0
P_{+j}	14.2	5.2	2.0	69.3	1.8	1.4	3.1	2.5	0.5	0.2	100	29.1	29.1	23.9	5.2		
Gain	7.7	1.4	1.3	12.8	0.9	0.7	3.0	1.1	0.3	0.0	29.1						

1994/2009	GL	AGL	SHB	CL	RF	WL	PF	ML	BL	EF	Pi+	Loss	C_i	S_i	D_i	l_p	g_p
GL	7.0		0.2	5.5	0.0	0.0	0.4	0.4	0.6		14.2	7.2	15.0	14.4	0.6	1.0	1.1
AGL	0.3	3.8		0.6	0.0	0.0	0.1	0.1	0.2	0.0	5.2	1.4	2.4	2.1	0.3	0.4	0.3
SHB	0.3	0.0	0.9	0.4		0.1	0.2	0.0	0.0	0.0	2.0	1.0	1.4	0.8	0.7	1.1	0.4
CL	5.8	0.8	0.0	60.1	0.1	0.2	1.1	1.0	0.1		69.3	9.1	18.5	18.3	0.3	0.2	0.2
RF	0.2	0.0		0.3	0.9	0.0	0.2	0.1	0.0	0.0	1.8	0.8	1.0	0.3	0.7	0.9	0.1
WL	0.2	0.0	0.1	0.2		0.6	0.2	0.1	0.0	0.0	1.4	0.8	1.4	1.4	0.1	1.3	1.1
PF	0.2	0.0	0.0	1.3	0.0	0.2	1.1	0.1	0.0		3.1	2.0	4.2	3.9	2.0	1.8	2.0
ML	0.8	0.1	0.0	1.0	0.0	0.0	0.1	0.4	0.0		2.5	2.1	3.9	3.8	0.2	4.9	4.5
BL	0.1	0.0	0.0	0.1	0.0	0.0	0.0	0.0	0.2		0.5	0.2	1.2	0.5	0.8	0.9	4.0
EF	0.0	0.0		0.0		0.0	0.0			0.2	0.2	0.0	0.1	0.0	0.0	0.2	0.1
P_{+j}	14.9	4.8	1.3	69.5	1.1	1.3	3.4	2.3	1.2	0.2	100	24.6	24.6	22.7	2.8		
Gain	7.8	1.0	0.4	9.4	0.1	0.7	2.2	1.9	1.0	0.0	24.6						

1957/2009	GL	AGL	SHB	CL	RF	WL	PF	ML	BL	EF	Pi+	Loss	C_i	S_i	D_i	l_p	g_p
GL	4.4	0.0	0.2	14.8	0.1	0.1	0.3	0.3	0.4	0.0	20.6	16.2	26.7	21.0	5.7	4.7	2.4
AGL	0.5	3.6	0.0	1.3	0.0	0.1	0.4	0.1	0.2	0.0	6.2	2.6	3.9	2.6	1.3	0.7	0.4
SHB	1.6	0.0	0.5	4.1	0.0	0.2	0.2	0.1	0.0	0.0	6.7	6.2	7.1	1.8	5.3	12.4	0.5
CL	5.4	0.4	0.2	43.8	0.2	0.3	1.4	1.2	0.5	0.0	53.4	9.6	35.3	19.2	16.1	0.2	0.6
RF	1.0	0.1	0.0	1.9	0.6	0.0	0.3	0.1	0.0	0.0	4.0	3.4	3.9	1.0	2.9	5.7	0.8
WL	0.6	0.0	0.4	2.0	0.0	0.5	0.3	0.1	0.0	0.0	3.9	3.4	4.2	1.6	2.6	6.8	1.6
PF	0.0	0.0	0.0	0.0	0.0	0.0	0.0	0.0	0.0	0.0	0.0	0.0	3.3	0.0	3.3		
ML	1.3	0.1	0.1	1.4	0.2	0.1	0.3	0.4	0.0	0.0	3.9	3.5	5.4	3.8	1.6	8.8	4.8
BL	0.1	0.0	0.0	0.2	0.0	0.0	0.1	0.0	0.0	0.0	0.4	0.4	1.5	0.8	0.7		
EF	0.0	0.7	0.0	0.0	0.0	0.0	0.0	0.0	0.0	0.2	0.9	0.7	0.7	0.0	0.7	3.5	0.0
P_{+j}	14.9	4.9	1.4	69.5	1.1	1.3	3.3	2.3	1.1	0.2	100	46.0	46.0	25.9	20.1		
Gain	10.5	1.3	0.9	25.7	0.5	0.8	3.3	1.9	1.1	0.0	46.0						

*The highlighted entries indicate persistence. P_{i+}= total time 1, P_{+j}= total time 2, C_j=total change, S_j=the amount of swap, D_j =the absolute value of the net change, lp =loss-to-persistence ratio, and g_p =gain-to-persistence

Higher annual rate of de-vegetation of shrubs and bushes (-6.3% a^{-1}) was observed during the period 1957-1972. *Ericaceous* forest was deforested at a higher annual rate during the periods 1972-1986 (-3.9% a^{-1}) and 1986-1994 (-9.7% a^{-1}) as compared to other classes. The annual average rates of deforestation of shrubs and bushes, riverine forest, wood land, and *Ericaceous* forest were 3.3%, 3%, 3%, and 4.1%, respectively, indicating *Ericaceous* forest was de-vegetated at a faster rate. The annual rate of afforestation/ reforestation (of plantation) increased during the periods 1972-1986 (21.4% a^{-1}) and 1986-1994 (18.3% a^{-1}). The recent (1994-2009) rates of deforestation of riverine forest (-4.44% a^{-1}) and shrubs and bushes (-3.69% a^{-1}) exceeded the rate of increase of recent (1994-2009) plantation forest (0.88% a^{-1}). This indicates that recent efforts of establishment of plantations are not balanced with recent deforestation. Therefore, in order to increase the environmental benefits from plantation forest, it is important to give due attention to plantation of selected forests. Figure 13c depicts the area affected by old deforestation (38.3 km^2) is much greater than that of the recent deforestation (7.4 km^2). The extent of stable forest (i.e. natural forest that remained unchanged over the past 52 years) accounted for only 2.2 km^2 (~1% of the landscape), which calls for a strong attention to conserve the remaining patches.

Table 7 Rate of change for different land use and land cover classes

Land use and land cover	1957 - 1972		1972 - 1986		1986 - 1994		1994 - 2009		Mean	
	km^{2}a^{-1}	% a^{-1}	km^{2}a^{-1}	% a^{-1}	km^{2}a^{-1}	% a^{-1}	km^{2}a^{-1}	% a^{-1}	km^{2}a^{-1}	% a^{-1}
GL	0.00	-0.2	-0.02	-1.7	-0.02	-1.9	0.00	0.45	-0.01	-0.8
AGL	0.00	-0.4	-0.01	-1.0	0.00	-0.2	-0.01	-0.55	-0.01	-0.5
SHB	-0.06	-6.3	-0.03	-2.6	-0.02	-2.1	-0.04	-3.69	-0.04	-3.7
CL	0.01	1.2	0.01	0.7	0.01	0.5	0.00	0.06	0.01	0.6
RF	-0.05	-5.5	0.00	-0.5	-0.02	-1.6	-0.04	-4.44	-0.03	-3.0
WL	-0.03	-2.6	-0.01	-0.7	-0.08	-8.3	-0.01	-0.51	-0.03	-3.0
PF	-	-	0.21	21.4	0.18	18.3	0.01	0.88	0.10	10.1
ML	-0.02	-1.8	-0.01	-1.3	-0.01	-1.0	-0.01	-0.63	-0.01	-1.2
BL	-0.01	-0.7	0.06	5.6	-0.08	-7.7	0.09	8.69	0.02	1.5
EF	-0.03	-2.7	-0.04	-3.9	-0.10	-9.7	0.00	0.00	-0.04	-4.1

3.5.3 Detection of most systematic transitions

The difference between observed and expected gains under a random process of change for grassland–cultivated land transition is 3.5% (Table 7 and Table 8). Thus, the transition of 15% of the landscape from grassland to cultivated land was due to systematic processes of change. The difference between observed and expected gains for afro-alpine grassland-cultivated land (-2.1%) is relatively large and the negative value suggesting that when cropland gains, a higher tendency of cultivated land to avoid gaining systematically from afro-alpine grassland. When grassland gains it tends to systematically gain from shrubs and bushes, riverine forest, and marshland but not from cultivated land. The difference between observed and expected losses under a random process of loss for grassland-cultivated land was 1.7%. Thus, there is a systematic transition from grassland to cultivated land as cultivated land was systematically gaining from grassland and at the same time grassland was also systematically losing to cultivated land. However, a systematic transition was not observed in the cultivated land to grassland change. Even though cultivated land was systematically losing to grassland (0.7%), grassland rather systematically avoided gaining from cultivated land (-1.6%). One of the implications of cultivated land expansion at the expense of grassland is scarcity of grazing land. Grazing land scarcity could induce resource use conflicts and stimulate unsustainable resource use practices such as overstocking that leads to overgrazing. On the one hand, grazing land scarcity will consequently bring increased land degradation due to overgrazing, and on the other hand it will lead to the decline in livestock population due to shortage of pastures. Therefore, a systematic transition of grassland to cultivated land contributes to the increase in poverty levels among agro-pastoral farmers of the Jedeb watershed.

Table 8 Transitions in percentage of total landscape under random process of gain (Jagtap et al.), and random process of loss (L_{ij}) for the period 1957-2009.

1957/2009	Variable	GL	AGL	SHB	CL	RF	WL	PF	ML	BL	EF	P$_{i+}$	Loss
GL	G_{ij}	4.4	0.3	0.2	11.4	0.1	0.2	0.7	0.4	0.2	0.0	17.8	13.4
	L_{ij}	4.4	0.9	0.3	13.2	0.2	0.2	0.6	0.4	0.2	0.0	20.6	16.2
AGL	G_{ij}	0.8	3.6	0.1	3.4	0.0	0.1	0.2	0.1	0.1	0.0	8.4	4.8
	L_{ij}	0.4	3.6	0.0	1.9	0.0	0.0	0.1	0.1	0.0	0.0	6.2	2.6
SHB	G_{ij}	0.9	0.1	0.5	3.7	0.0	0.1	0.2	0.1	0.1	0.0	5.7	5.2
	L_{ij}	0.9	0.3	0.5	4.4	0.1	0.1	0.2	0.1	0.1	0.0	6.7	6.2
CL	G_{ij}	7.1	0.7	0.5	43.8	0.3	0.4	1.8	1.1	0.6	0.0	56.2	12.4
	L_{ij}	4.7	1.5	0.4	43.8	0.3	0.4	1.0	0.7	0.3	0.1	53.4	9.6
RF	G_{ij}	0.5	0.1	0.0	2.2	0.6	0.0	0.1	0.1	0.0	0.0	3.7	3.1
	L_{ij}	0.5	0.2	0.0	2.4	0.6	0.0	0.1	0.1	0.0	0.0	4.0	3.4
WL	G_{ij}	0.5	0.1	0.0	2.2	0.0	0.5	0.1	0.1	0.0	0.0	3.5	3.0
	L_{ij}	0.5	0.2	0.0	2.4	0.0	0.5	0.1	0.1	0.0	0.0	3.9	3.4
PF	G_{ij}	0.0	0.0	0.0	0.0	0.0	0.0	0.0	0.0	0.0	0.0	0.0	0.0
	L_{ij}	0.0	0.0	0.0	0.0	0.0	0.0	0.0	0.0	0.0	0.0	0.0	0.0
ML	G_{ij}	0.5	0.1	0.0	2.2	0.0	0.0	0.1	0.4	0.0	0.0	3.4	3.0
	L_{ij}	0.5	0.2	0.1	2.5	0.0	0.0	0.1	0.4	0.0	0.0	3.9	3.5
BL	G_{ij}	0.1	0.0	0.0	0.2	0.0	0.0	0.0	0.0	0.0	0.0	0.3	0.3
	L_{ij}	0.1	0.0	0.0	0.3	0.0	0.0	0.0	0.0	0.0	0.0	0.4	0.4
EF	G_{ij}	0.1	0.0	0.0	0.5	0.0	0.0	0.0	0.0	0.0	0.2	0.9	0.7
	L_{ij}	0.1	0.0	0.0	0.5	0.0	0.0	0.0	0.0	0.0	0.2	0.9	0.7
P$_{i+}$	G_{ij}	14.9	4.9	1.4	69.5	1.1	1.3	3.3	2.3	1.1	0.2	100	46.0
	L_{ij}	12.2	7.0	1.4	71.3	1.3	1.4	2.3	2.0	0.8	0.3	100	46.0
Gain	G_{ij}	10.5	1.3	0.9	25.7	0.5	0.8	3.3	1.9	1.1	0.0	46.0	
	L_{ij}	7.8	3.4	0.9	27.5	0.7	0.9	2.3	1.6	0.8	0.1	46.0	

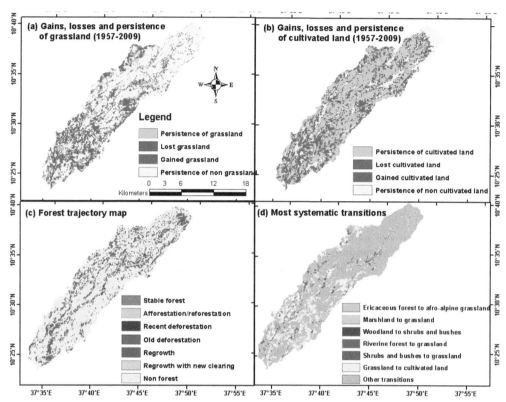

Figure 13 Spatial representation of the gains, losses, and persistence experienced in (a) grassland and (b) cultivated land, (c) forest trajectory, and (d) most systematic land use and land cover transitions for the period 1957–2009

The difference between observed and expected losses for grassland-plantation forest, afro-alpine grassland-plantation forest, riverine forest-plantation forest, wood land-plantation forest and marshland-plantation forest transitions were -0.3%, 0.3%, 0.2%, 0.2%, and 0.2%, respectively (Table 7 and Table 8). These results indicate that plantation forest avoided gaining from grassland systematically but it tended to replace afro-alpine grassland, riverine forest, wood land and marshland systematically. This pattern is attributable to the establishment of *Eucalyptus globules* and *Eucalyptus camaldulensis* plantations mainly for wood for fuel and construction uses (Getahun, 2002). During the military regime (1974-1991) *Eucalyptus globules* plantations were established through government supported afforestation programs, funded in some areas partly by international organizations (Zeleke and Hurni, 2001). During the current regime (1991- to date) increased market demand and attractive price of *Eucalyptus* poles that resulted from urbanization, road development and human population pressure have acted as driving forces to the establishment of *Eucalyptus* forest plantations. Bewket (2003) demonstrated the fact that the major factor in increased area coverage of forests has since recently been planting of trees at the household level.

When cultivated land gains, it is inclined to gain only from grassland systematically and disinclined to gain from the other categories. This result indicates that farmers in the Jedeb watershed are inclined to convert grassland rather than natural forest into cultivation, partly because of the ease with which grasslands could be converted to cultivated lands in terms of the required land clearing and the fact that forests are generally located in rough terrain and steep slopes that are unsuitable for cultivation. Therefore, the widely held view that expansion of agriculture is the primary cause for the loss of natural woody vegetation was not found to hold true in the case of Jedeb watershed. The loss of natural woody vegetation is largely attributed to increased demand for wood for fuel, farm implements, construction and other uses. The observed pattern, in the case of Jedeb watershed, was conversion of natural woody vegetation initially to grassland and subsequently grassland into cultivated land. These findings agree with those of Tegene (2002) who indicated for his study area that expansion of cultivation is not the primary cause for disappearance of woody vegetation.

Based on the above analyses, the most dominant signals of changes in the Jedeb watershed during the period 1957-2009 were: (1) conversion of about 14.8% of grassland to cultivated land, (2) the conversion of shrubs and bushes, riverine forest, and marshland to grassland which accounts for about 3.9% of the landscape, (3) the conversion of *Ericaceous* forest to afro-alpine grassland which accounts for 0.7% of the landscape, and (4) the conversion of wood land to shrubs and bushes which accounts for 0.4% of the landscape.

Table 9 Percent of land use and land cover changes in terms of gains and losses for the period 1957-2009*

1957/2009	GL		AGL		SHB		CL		RF		WL		PF		ML		BL		EF	
	D_{ij}	R_{ij}	D_{ij}	R_{ij}	D_{ij}	R_{ij}	D_{ij}	R_{ij}	D_{ij}	R_{ij}	D_{ij}	R_{ij}	D_{ij}	R_{ij}	D_{ij}	R_{ij}	D_{ij}	R_{ij}	D_{ij}	R_{ij}
GL	0.0	0.0	-0.3	-1.0	0.0	0.0	3.4	0.3	0.0	-0.1	-0.1	-0.4	-0.4	-0.6	-0.1	-0.3	0.2	0.8	0.0	
	0.0	0.0	-0.9	-1.0	-0.1	-0.2	1.6	0.1	-0.1	-0.5	-0.1	-0.6	-0.3	-0.5	-0.1	-0.3	0.2	0.9	0.0	-1.0
AGL	-0.3	-0.4	0.0	0.0	-0.1	-1.0	-2.1	-0.6	0.0	-1.0	0.0	0.9	0.2	1.0	0.0	-0.2	0.1	1.9	0.0	
	0.1	0.2	0.0	0.0	0.0	-1.0	-0.6	-0.3	0.0	-1.0	0.1	1.8	0.3	3.4	0.0	0.6	0.2	5.7	0.0	-1.0
SHB	0.7	0.8	-0.1	-1.0	0.0	0.0	0.4	0.1	0.0	-1.0	0.1	2.6	0.0	-0.1	0.0	-0.2	-0.1	-1.0	0.0	
	0.7	0.7	-0.3	-1.0	0.0	0.0	-0.3	-0.1	-0.1	-1.0	0.1	1.4	0.0	0.0	0.0	-0.3	-0.1	-1.0	0.0	-1.0
CL	-1.7	-0.2	-0.3	-0.5	-0.3	-0.6	0.0	0.0	-0.1	-0.3	-0.1	-0.3	-0.4	-0.2	0.1	0.1	-0.1	-0.2	0.0	
	0.7	0.2	-1.1	-0.7	-0.2	-0.5	0.0	0.0	-0.1	-0.4	-0.1	-0.3	0.4	0.3	0.5	0.7	0.2	0.4	-0.1	-1.0
RF	0.5	0.9	0.0	0.8	0.0	-1.0	-0.3	-0.1	0.0	0.0	0.0	-1.0	0.2	1.3	0.0	0.3	0.0	-1.0	0.0	
	0.5	1.0	-0.1	-0.4	-0.5	-1.0	-0.5	-0.2	0.0	0.0	0.0	-1.0	0.2	1.6	0.0	0.3	0.0	-1.0	0.0	-1.0
WL	0.1	0.2	-0.1	-1.0	0.4	9.6	-0.2	-0.1	0.0	-1.0	0.0	0.0	0.2	1.3	0.0	0.3	0.0	-1.0	0.0	
	0.1	0.2	-0.2	-1.0	0.4	7.3	-0.4	-0.2	0.0	-1.0	0.0	0.0	0.2	1.6	0.0	0.3	0.0	-1.0	0.0	-1.0
PF	0.0		0.0		0.0		0.0		0.0		0.0		0.0		0.0		0.0		0.0	
	0.0		0.0		0.0		0.0		0.0		0.0		0.0		0.0		0.0		0.0	
ML	0.8	1.5	0.0	0.9	0.1	1.7	-0.8	-0.3	0.2	8.8	0.1	2.1	0.2	1.3	0.0	0.0	0.0	-1.0	0.0	
	0.8	1.4	-0.1	-0.4	0.0	1.0	-1.1	-0.4	0.2	4.1	0.1	1.1	0.2	1.5	0.0	0.0	0.0	-1.0	0.0	-1.0
BL	0.0	0.9	0.0	-1.0	0.0	-1.0	0.0	-0.1	0.0	-1.0	0.0	-1.0	0.1	6.6	0.0	-1.0	0.0		0.0	
	0.0	0.7	0.0	-1.0	-0.1	-1.0	-0.1	-0.3	0.0	-1.0	0.0	-1.0	0.1	6.5	0.0	-1.0	0.0		0.0	-1.0
EF	-0.1	-1.0	0.7	55.1	0.0	-1.0	-0.5	-1.0	0.0	-1.0	0.0	-1.0	0.0	-1.0	0.0	-1.0	-1.0	-1.0	0.0	0.0
	-0.1	-1.0	0.7	19.4	0.0	-1.0	-0.5	-1.0	0.0	-1.0	0.0	-1.0	0.0	-1.0	0.0	-1.0	-1.0	-1.0	0.0	0.0

*D_{ij} = the difference between the observed and the expected value, R_{ij} = the difference between the observed and the expected value, relative to the expected value. The numbers in bold are values for the gains (%) and the numbers in normal font are values for the losses (%). The most systematic transitions are highlighted.

3.5.4 Derminants of locations of the most systematic transitions

Forest-to-grassland conversion

A binary logistic regression analysis was conducted to analyse the factors determining the patterns of forest-to-grassland conversion using 11 predictors (Table 9) for the period 1957-2009. A test of the full model against a constant only model was statistically significant, indicating that the predictors as a set reliably distinguished forest to grassland conversation against all other conversions (χ^2 (11, N=764) = 500.47, $p < 0.000$). The model as a whole explained 64.1% (Nagelkerke pseudo R^2) of the variance, and correctly classified 86.8% of the cases. The ROC value of this model is 0.91(Table 10), indicating that the goodness of fit of the model is excellent according to the rating explained by Hosmer and Lemeshow (2000). The result of this spatially explicit logistic regression model is illustrated in Figure 14d. As shown in Table 10, *travel time to town (market)* and *north-facing slope (cosine of aspect)* did not have any statistically significant contribution to forest-to-grassland conversion.

Table 10 Binary logistic regression results for grassland-to-cultivated land conversion and forest-to-grassland conversion

Dependent variable	Model evaluation	Independent variables	Unstandardized Logit Coefficient (β)	Standard error of β	Standardized Logit Coefficient (β^*)
GL-CL	LR χ^2=389.29	SLP	-0.0613	0.0113	-0.3420
Conversion	(p=0.000)	ELEV	-0.0032	0.0003	-1.0180
(1957/2009)		SIN_ASP	*	*	*
	Nagelkerke	COS_ASP	*	*	*
	R^2=0.25	TWI	-0.1135	0.0279	-0.2370
		D_RIVER	*	*	*
		TT_ROAD	0.0065	0.0020	0.2130
	ROC=0.72	TT_TOWN	-0.0037	0.0015	-0.1730
		PD57	-0.0071	0.0021	-0.2300
		PD07_57	-0.0088	0.0020	-0.2590
		Intercept	10.6446	0.8178	-0.1750
F-GL	LR χ^2=500.47	SLP	0.0475	0.0187	0.302
Conversion	(p=0.000)	ELEV	0.0019	0.0005	0.971
(1957/2009)		SIN_ASP	0.8679	0.1647	0.624
		COS_ASP	*	*	*
	Nagelkerke	TWI	0.2680	0.0607	0.572
	R^2=0.64	D_FOREDGE57	-0.0163	0.0016	-4.484
		D_RIVER	-0.0007	0.0003	-0.408
		TT_ROAD	-0.0111	0.0043	-0.353
	ROC=0.91	TT_TOWN	*	*	*
		PD57	-0.0123	0.0041	-0.410
		PD07_57	*	*	*
		Intercept	-5.6701	1.3102	-1.231

* Not significant, all other values are significant at 0.05 level
Note: The standardized logit coefficient is the unstandardized logit coefficient multiplied by the parameter's standard deviation.
LR χ^2: Likelihood Ratio or Model Chi-square. ROC: Relative Operating Characteristics

Topographic variables: The variables elevation, sine of aspect (eastness), topographic wetness index and slope appear to be positively related with forest-to-grassland conversion. Thus, the likelihood of forest-to-grassland conversion increases with increasing elevation, eastness, soil wetness, and steepness of a location. A unit standard deviation increase in elevation (488 m) is associated with a 0.971 standard deviation increase in the logit of forest-to-grassland conversion. This is to be explained by the existence of high livestock density in the high elevation areas (e.g. *Sinan* wereda = 138 $\#/km^2$) than in the lower elevation areas (e.g. *Mchakel* wereda = 108 $\#/km^2$) of Jedeb watershed (Table 11). Thus, cattle grazing could be the immediate cause for forest-to-grassland conversion. A unit standard deviation increase in slope (6.4°) is associated with a 0.302 standard deviation increase in the logit of forest-to-grassland conversion, suggesting forest areas with steep slopes are more likely to be converted to grassland.

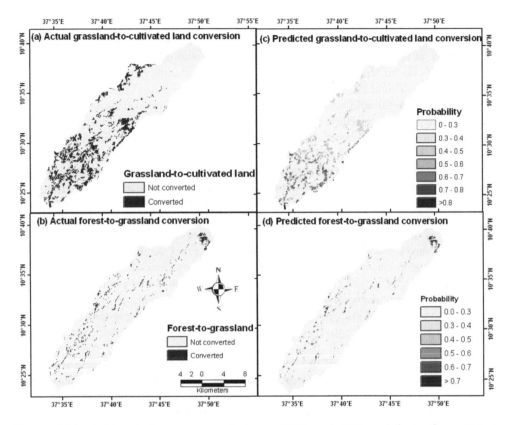

Figure 14 (a & b) actual conversions between 1957 and 2009 and (c & d) predicted conversions based on the logistic regression models in Table 10

Table 11 Human (rural) and livestock population density in the two *Kebelles* (districts) of Jedeb watershed

Wereda (District)	Mean elevation (m)*	Population density (#/km²)	
		Human (Rural)**	Livestock***
Machakel	2284	148.48	107.62
Sinan	2900	216.24	137.49

Source: *SRTM90   ** CSA, 2007   *** Office East Gojam Zone Agriculture and Rural Development (unpublished report, 2010)

Distance variables: All statistically significant distance variables are negatively correlated with forest-to-grassland conversion. Thus, deforestation decreases with increasing *distance from forest edges, rivers* and *roads*. *Distance from forest edge* appears to have the strongest effect (-4.484 in Table 10) in explaining forest-to-grassland conversion, followed by elevation, slope aspect, wetness, population density, proximity to river, proximity to road, and then slope. The probability of forest-to-grassland conversion decreases from 0.68 to 0.31 for an increase in *distance from forest edge* of 0-120m. Even at a distance of 60 m from forest edges, the predicted probability is 0.49. This suggests that not only at the edges of forest but also at a distance of 60 m from the edge there is a moderately higher probability of deforestation. Therefore, a 240 m-buffer of existing forest can protect it from conversion to grassland, because the predicted probability at this distance is 0.08 (Fig. 15e). A unit standard deviation decrease in *distance from river* (537 m) is associated with a 0.408 standard deviation increase in the logit of forest-to-grassland conversion. Forests located near rivers and on wet soils (higher *topographic wetness index*) seem to have higher likelihood of being converted to grassland. This indicates the susceptibility of riverine forest to deforestation than any other forest types in the Jedeb watershed. This finding is in line with results reported from elsewhere at local scale analysis of deforestation (e.g.Wyman and Stein, 2010). *Travel time to town (market)* is not a statistically significant predictor of the location of forest-to-grassland conversion. This implies that the local market is not influencing forest-to-grassland conversion as most of the forest products are instead transported to other larger towns including the capital city, Addis Ababa. The model predicted that a unit standard deviation decrease in *travel time to road* (31.7 min) is associated with a 0.353 standard deviation increase in the logit of forest-to-grassland conversion, indicating that forests located near roads are more likely to be deforested as lower travel costs increase returns from forests. At zero *travel time to road* the model predicted a 0.28 likelihood of deforestation. This indicates *travel time to road* has a moderate power in explaining forest-to-grassland conversion. If forest conservation planners want to minimize the effect of roads on deforestation it is possible to delineate areas for afforestation at beyond 80 min *travel time from roads* (Fig. 15b).

Population density: The Jedeb watershed was inhabited by a total population of about 47,448 in 2007, growing at a rate of 2.02% per annum. Assuming that this rate of growth has remained constant, the total population of the watershed would have been around 17,281 in 1957. Similarly, the estimated population for the year 2012 would be 52,490 and the population will be double within 35 years from now (Table 12). Despite the rise in population pressure over the past 50 years, an increase in population density between 1957 and 2007 was not a significant determinant of forest-to-grassland conversion (Table 10). The population potential surface derived from *kebelle* population acquired from CSA (2008) was not found to be a significant predictor of forest-to-grassland conversion. The model predicts that a unit standard deviation increase in *initial population density* (32 persons/km^2) is associated with a 0.41 standard deviation decrease in the logit of forest-to-grassland conversion. This could raise several hypotheses, as it is difficult to determine causal relationships. This could be related to the establishment of *Eucalyptus* plantations around homesteads in place of deforested riverine forest and woodland as a result of population pressure. This result accords with some earlier analysis (e.g.Leach and Fairhead, 2000; Bewket, 2003) who explored neo-Boserupian perspective of population-forest dynamics.

Table 12 Demographic characteristics of Jedeb Watershed*

Demographic variables	1957 [a]	2007[b]
Total population size	1 721 (15 909)	47 448 (29 115)
Number of households	3 677 (3 387)	10 612 (6 470)
Average population density (Persons/km^2)	58 (50)	160 (91)

*Numbers in parentheses are estimations based on counting the number of huts ("tukul") using 4.7 and 4.5 average inhabitants for 1957 and 2007, respectively.
[a]The projection is based on the assumption of exponential increase: $P_n = P_o e^{r*n}$
[b]Source: CSA, 2008

Grassland-to-cultivated land conversion
Another binary logistic regression analysis was conducted to predict grassland-to-cropland conversion using ten predictors for the period 1957-2009. A test of the full model against a constant only model was statistically significant, indicating that the predictors as a set reliably distinguished grassland-to-cropland conversion against all other conversions (χ^2 (10, N=1876) =389.29, $p < 0.000$). The model as a whole explained 25% (Nagelkerke pseudo R^2) of the variance, and correctly classified 71.2% of the cases. The result of this spatially explicit logistic regression model is illustrated in Figure 14c. The ROC value of 0.72 (Table 10) suggests that the model fit is generally good. Proximity to river and slope aspect did not have any statistically significant contribution to grassland-to-cropland conversion (Table 10). Grassland located near rivers seems to have little likelihood of being converted to cultivated land even though there is a good possibility of getting water for irrigation.

Topographic variables: Based on the standardized logistic regression coefficients, *elevation* appears to have the strongest explanatory power, followed by *slope*. A unit standard deviation decrease in elevation (312 m) is associated with a 1.018 standard deviation increase in the logit of grassland-cultivated land conversion. Similarly, a unit standard deviation decrease in slope (5.4°) is associated with a 0.342 standard deviation increase in the logit of grassland-to- cropland conversion. Topographic wetness index has a statistically significant negative relationship with grassland-cultivated land conversion. A unit standard deviation decrease in *topographic wetness index* is associated with a 0.237 standard deviation increase in the logit of grassland-cultivated land conversion. This indicates that farmers do not take soil wetness into consideration when they tend to convert grassland to cultivated land. The above results indicate only two of the geo-physical determinants of agricultural potential (elevation and slope) reinforce the Ricardian land change theory.

Distance variable: Travel time to market (Townshend and Justice) is another important determinant of the location of grassland-to-cropland conversion. A unit standard deviation decrease in travel time (46.6 min) is associated with a 0.173 standard deviation increase in the logit of grassland-to-cropland conversion, suggesting that the shorter the travel time to markets the higher is the likelihood of grassland-to-cropland conversion. This is because the reduced cost to transport inputs and outputs to the market as a result of shorter travel time may stimulate increases in agricultural production. This finding very well agrees with that of Dercon and

Hoddinott (2005), who explained the more remote households are from towns, the less likely they are to purchase inputs or sell a variety of products in Ethiopia. This result clearly indicates the land use pattern in the Jedeb watershed is shaped by von Thunen's theory. However, *travel time to road* is positively correlated with the grassland-to-cultivated land conversion, suggesting that local people do not use existing roads intensively to come to markets/towns. This could be explained by poor access to and/or high cost of transportation for the local people.

Population pressure: Low population density in 1957 increased the likelihood of observing grassland-to-cultivated land conversion, suggesting that low population density is an important determinant of agricultural extensification. The model predicts that a unit standard deviation decrease in *initial population density* (32 persons/km^2) is associated with a 0.23 standard deviation increase in the logit of grassland-to-cultivated land conversion (Table 10). This finding is in line with that of Braimoh and Vlek (2005) who indicated that population density is important determinant where the dominant change process is extensification. An increase in population density between 1957 and 2007 significantly decreased the probability of observing the grassland-to-cultivated land conversion. This result suggests that an increasing population pressure could rather lead to intensified cultivation of land as a land use change process than an agricultural extensification.

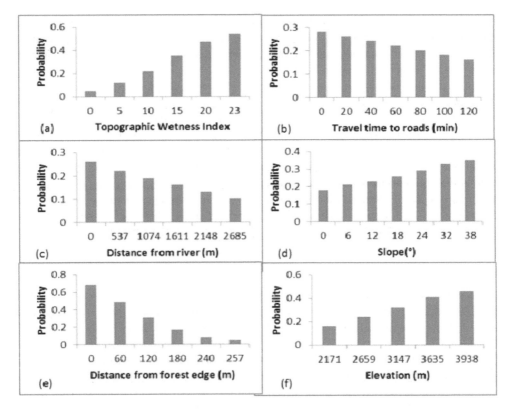

Figure 15 Probability of forest-to-grassland conversion according to different explanatory variables

3.6   Conclusions

Analysis of multi-temporal LULCC through enhanced transition matrix and spatial statistical modelling improved the identification, quantification and understanding of determinants of most systematic transitions in the Jedeb watershed, a headwater tributary of the Blue Nile River in Ethiopia. The study found that the watershed has undergone significant land use/ cover alterations since 1957. In all of the time spans considered (1957-1972, 1972-1986, 1986-1994, 1994-2009), cultivated land and grassland constituted the predominant types of land cover, although the former has been by far the largest coverage. Of the total area of the watershed, cropland accounted for 53.4% in 1957, 61.8% in 1972, 66.5% in 1986, 69.3% in 1994 and 69.5% in 2009. A clear trend of growth in cultivated land until 1994 was found. The absence of a significant increase in the cultivated land since 1994 suggests that all the area that is suitable for cultivation has most likely been used, or the land tenure system has effectively controlled spontaneous expansion of cultivation by local people.

About 46% of the watershed area experienced transition from one category to a different category of land use/ cover over the 52 years considered. Out of the 46%, about 20% of the changed area was a net change while 25.9% was a swap change. Swap change is greater than net change suggesting the importance of the swapping component and common methods of land use/ cover change study would miss this dynamics. Cultivated land and afro-alpine grassland tend to persist; riverine forest and shrubs and bushes tend to lose; and grassland, woodland and marshland tend to gain or lose rather than persist. The type of change that each land use/ cover experienced differs from period to period. For example, when the difference between gains and losses experienced by grassland is small (1957-1972 and 1994-2009), grassland experienced mainly swap type of change. Otherwise, it experienced both net and swap types of change. Identification of swapping change dynamics is important especially in the absence of the net change which may otherwise be interpreted as absence of change because the net change fails to capture the swapping component of the change.

The rate of afforestation/ reforestation (plantation) far outpaced that of deforestation in the periods 1972-1986 and 1986-1994, whereas recent (1994-2009) rates of deforestation of riverine forest (-4.44% a^{-1}) and shrubs and bushes (-3.69% a^{-1}) exceeded the rate of increase of recent (1994-2009) rate plantation forest (0.88% a^{-1}). Plantation forest grew at the expense of afro-alpine grassland, riverine forest, wood land and marshland systematically. Increasing demands and attractive prices of *Eucalyptus* poles resulting from urbanization, road development and human population pressure are the main driving forces for the recent expansion of forest plantations.

Lower elevations, gentle slopes, less populated areas, locations near to markets/towns, and locations farther from roads increase the likelihood of grassland-to-cultivated land conversion in the Jedeb watershed. Farmers of the Jedeb watershed are inclined to convert grassland rather than natural forest for cultivation. The observed overall pattern has been conversion of natural woody vegetation to grassland and then grassland was subsequently converted to cultivated land. Therefore, the widely held view that expansion of agriculture is the primary cause for the loss of natural woody vegetation in Ethiopia was not found to hold true in the case of the Jedeb watershed. Forest edges, higher elevations, east-facing and steeper slopes, less populated areas, locations near roads and rivers, and locations with high soil wetness increase the likelihood of forest-to-grassland conversion. A 240 m-buffer of existing forest is likely to protect forest edges from conversion to grassland. Forests located near rivers and on wet soil (i.e. riverine forests) seem to have higher susceptibility to deforestation than the other forest types. Forests located near roads are more likely to be deforested, as lower travel costs increase returns from forests. Thus, the loss of forest could be largely due to increased demand for wood for fuel, construction, farm implements and other uses. The spatial statistical modelling of most systematic transitions in this study suggests that spatial determinants of LULCC can be related to the well-established land change theories. The determinants of grassland-to-cultivated conversion model such as *travel time to market* reinforce evidence of von Thunen's model, whilst the geo-physical determinants of agricultural potential (elevation and slope) reinforce the Ricardian land change theory.

Finally, this study has highlighted that integrated use of Remote Sensing and GIS technologies improves quantification; statistical modelling and therefore improved understanding of the process of land use/ cover change. Thus, identifying most systematic transitions and its spatial statistical modelling under GIS environment provides essential planning tools for sustainable watershed management.

Chapter 4

SPATIO-TEMPORAL ANALYSIS OF CHANGES IN

LANDSCAPE PATTERNS IN THE JEDEB

WATERSHED DURING THE PERIOD 1957-2009[2]

4.1 ABSTRACT

Different landscape metrics indicated that the spatio-temporal pattern of the Jedeb landscape has undergone significant alterations in terms of composition and configuration since 1957. Landscape fragmentation has increased during the period 1957 to 2009 as discerned by the increase in number of smaller patches (an increase in the number of patches (NP) from 1621 to 5179 and a decrease in the mean patch size (MPS) from 18.3 to 5.7 ha). Patch size relative variability increased over time (1743 to 4933 ha) suggesting that landscape heterogeneity increased posing a land management challenge. In the year 2009 the Jedeb watershed was dominated by a few contiguous patches and all patch types were aggregated as the contagion index (CONTAG) has increased from 57% to 65% during the period 1957-2009. This increase in dominance is also supported by the increase in the largest patch index (LPI) from 42% to 68% and a decrease in the Shanon's Diversity Index (SHDI) value from 1.47 to 1.11 in 1957 and 2009, respectively. The mean shape index (MSI) ranked the Jedeb landscape along an intuitive temporal gradient from most complex to simple shapes (1.57 to 1.24). The patches were most irregular and complex in shape

---

[2]*This chapter is based on* Teferi, E., Bewket, W., Uhlenbrook, S., Wenninger, J., (2015).Spatio-temporal analysis of landscape patterns in the Jedeb watershed during the period 1957-2009 *(In preparation)*.

.

during the year 1957, whereas patches were becoming least irregular and simple in shape progressively from 1957 onwards. These results indicate that fragmentation is causing a simplification in patch shape compared to the geometrically complex patch shapes found in the past.

4.2 INTRODUCTION

Changes in a landscape are resulted from the complex interactions between physical, biological, economic, political and social factors. Understanding the complexity of landscapes require mapping and quantifying the landscape pattern through time (Narumalani et al., 2004). Landscape pattern implies spatial heterogeneity. Spatial heterogeneity has two components: the amounts of different possible entities (composition), and their spatial arrangements (configuration). Investigating changes in landscape pattern is a prerequisite to the study of ecosystem functions and processes, sustainable resource management, and effective land use planning (Matsushita et al., 2006). Furthermore, in order to understand the causes and consequences of the patterns of land use/land cover change, quantifying landscape pattern is very crucial (Turner, 2005).

Landscape pattern analysis requires quantification of landscape pattern metrics (Gautam et al., 2003; Wu et al., 2000). Landscape metrics are algorithms that quantify specific spatial characteristics of patches, classes of patches, or entire landscape mosaics and provide ecologists and resource managers with a suite of tools for effective management decisions in conservation and planning (Ferreira et al., 2006). For example, high number of patch and low mean patch size (MPS) serve as important indicators of fragmented landscape condition (Matsushita et al., 2006; Southworth et al., 2004).

Despite several research efforts on land use and land cover changes (LULCC) in Ethiopia (Amsalu et al., 2007; Bewket, 2002; Tegene, 2002; Tsegaye et al., 2010; Zeleke and Hurni, 2001), landscape pattern analysis was not considered so far. An assessment of landscape patterns can help identify some of the most important aspects of environmental changes, which emerge from lower-level disturbances due to complex interactions between social and environmental processes (Forman, 1995; Turner, 1989). Thus, there is a need for examining landscape pattern and its changes in the source region of UBN river basin by using remote sensing and landscape metrics. The objectives of this study were to quantify landscape patterns and the changes in landscape patterns over time.

4.3   Data and Methods

4.3.1   Study Area

The study area, Jedeb watershed, lies between 10°23' to 10°40' N and 37°33' to 37°60' E. It is situated in the south-western part of Mount Choke, which is a headwater area of the Upper Blue Nile/Abay in Ethiopia (Fig. 6). Detailed description of the study area can be found in Section 3.3.

4.3.2   Data used

Black and white aerial photographs of 1957 and Landsat imageries of 1972 (MSS), 1986 (TM), 1994 (TM) and 2009 (TM) were used to derive ten land use and land cover classes by integrated use of Remote Sensing (RS) and Geographic Information System. The detailed procedure for extraction of land cover classes can be found in Chapter 3.

4.4   Quantifying landscape patterns

Prior to quantifying landscape metrics, the following procedures were implemented. These include defining the landscape boundary, establishing a model of landscape structure, and establishing a relevant grain of analysis and data model. The landscape boundary was defined based on a watershed as it is the basis for land management and future assessment of hydrological impacts. The patch mosaic model (also called patch-corridor-matrix model) of landscape structure was chosen, because the landscape analysis software (FRAGSTAT 3.3) used in this study is based on this framework. A relevant grain of analysis and data format was established. In this study a grain size of 30 m was chosen due to the resolution of some of the source data (e.g., 30 m resolution of Landsat imagery). Therefore, patches as small as 0.09 ha (1 cell) can be depicted.

The spatial pattern of land use/land cover (composition and configuration) in the Jedeb landscape was quantified using selected landscape metrics at the class as well as landscape level with FRAGSTAT 3.3, spatial pattern analysis program for categorical maps (McGarigal, 2002). Although FRAGSTAT computes more than 60 landscape metrics, many of them are redundant or inter-correlated (Frohn and Hao, 2006; Riitters et al., 1995). Several studies have attempted to define core set of independent metrics based on theoretical considerations and research objectives (Aguilera et al., 2011; Leitao and Ahern, 2002; Neel et al., 2004). Others have used statistical data reduction techniques to define an optimal set of metrics (Cushman et al., 2008; Riitters et al., 1995; Schindler et al., 2008). Although these studies recommend that patterns can be characterized by only a parsimonious suite of landscape metrics, consensus does not exist on the choice of individual metrics (Leitão et al., 2012). In this research 15 landscape metrics were chosen (Table 13) on the basis of theoretical considerations, analysis of literature, and relevance to this research. The selected metrics are used to characterize the spatial heterogeneity, fragmentation, complexity of patch shape, and connectivity for a given landscape.

Table 13 Selected landscape metrics (McGarigal, 2002)

Acronym	Metric Name (units)	Description
PLAND	Percentage of landscape (percent)	A relative measure of the proportional abundance of each patch type in the landscape; a measure of landscape composition.
NP	Number of patches (-)	The number of patches in the landscape or class; reflecting spatial heterogeneity and fragmentation.
LPI	Largest patch index (percent)	The percentage of total landscape area comprised by the largest patch; measure of dominance.
ED	Edge density (meters per hectares)	Higher values indicate more edge habitats, more fragmentation.
MPS	Mean patch size (hectares)	The arithmetic average size of each patches; reflects fragmentation.
MPSCV	Patch size coefficient of variation (percent)	Measures relative variability about the mean; indicates landscape heterogeneity.
MSI	Mean shape index (-)	A measure of the subdivision of the class or landscape.
AWMSI	Area-weighted mean shape index (-)	Measures the average patch shape weighted by patch area; index of shape complexity.
CONTAG	Contagion (percent)	Gives the magnitude of overall clumpiness as a percentage of the Maximum; it also indicates diversity and dominance.
CLUMPY	Clumpiness index (-)	Provides a measure of class-specific contagion that is not confounded by changes in class area; index of fragmentation.
CONNECT	Connectance index (percent)	The number of functional joining between patches of the same type.
PROX_MN	Mean proximity index (-)	A measure of patch isolation.
ENN_MN	Mean Euclidean Nearest Neighbour Distance (m)	The shortest distance from one patch to another patch of the same land covers type.
MSIDI	Simpson's Diversity Index (-)	Represents the probability that any two pixels selected at random would be different patch types; a measure of landscape composition.
MSIEI	Shannon's evenness index (-)	Measures the evenness in distribution of area among patch types.

4.5   Results and Discussion

4.5.1   Changes in landscape pattern at landscape level

Table 14 illustrates the change in landscape composition and configuration (i.e. spatial heterogeneity) of the Jedeb watershed at a landscape level. The contagion index (CONTAG) value indicates how the overall degree of clumpiness of the Jedeb watershed has changed over time as cropland expansion has progressed. CONTAG increased progressively from 57% to 65% of the maximum aggregation during the period 1957-2009, suggesting that the landscape has changed from a more diverse to a less diverse arrangement of land cover. This is because a single class (i.e. cultivated land) occupied large percentage of the landscape. This increase in dominance is also supported by the increase in largest patch index (LPI) from 42% to 68%.

The information obtained from the contagion metric is more important in the initial characterization of a landscape for the purpose of landscape planning (Hunziker and Kienast, 1999). The contagion metric can make landscape planners aware of what the most useful subsequent measures might be. For example, the higher values of contagion observed in 2009 can be indirect indicator low levels of shared edge between different ecosystems or land cover types. Thus, edge metrics could be more effective subsequent measurements on this landscape. However, the contagion metric doesn't give new information without prior information about LPI. The increase in LPI over time could imply a decrease in spatial diversity. This is evidenced by the observed decrease in modified Simpson's diversity index (MSIDI) value from 1.08 to 0.67 in 1957 and 2009, respectively. Such changes in the MSIDI values over time are reflections of changes in ecological systems such as plant species diversity (Kie et al 2002; Honnay et al., 2003). Kumar et al. (2006) reported that both native and non-native plant species richness were significantly ($p < 0.05$) and positively correlated with Simpson's Diversity Index. Thus, the decreasing trend in MSIDI values may indicates a decreasing trend in species richness over the landscape.

Table 14 Landscape metrics change at landscape level for the period 1957-2009. The abbreviations are described in Table 13.

Year	NP	LPI	ED	MPS	MPSCV	MSI	AWMSI	CONTAG	CONNECT	MSIDI	MSIEI
1957	1621	42.27	60.87	18.29	1743	1.57	10.10	57.09	0.46	1.08	0.49
1972	4382	59.67	90.76	6.77	3958	1.32	21.81	61.25	0.21	0.84	0.37
1986	6877	63.73	96.57	4.31	5299	1.24	23.32	61.92	0.19	0.75	0.32
1994	5835	67.58	84.83	5.08	5173	1.24	22.18	64.45	0.18	0.68	0.30
2009	5179	68.45	79.90	5.73	4934	1.26	20.65	65.27	0.17	0.67	0.29

The number of patches showed a substantial increase from 1621 in the year 1957 to 5176 in the year 2009in the period 1986-2009 (Fig. 16). The observed general increase in number of patches is a concrete manifestation of less connectivity and greater isolation in the entire Jedeb landscape mosaic during the study period 1957-2009. NP should be used in conjugation with MPS since high NP and low MPS values provide a strong evidence of a fragmented landscape as indicated by previous studies (BotequilhaLeit˜ao and Ahern, 2002; Kohn and Walsb, 1994; Nilsson and Ouborg, 1993). For example, the larger number of patches (NP=6877) along with smaller mean patch size (MPS=4.31 ha) observed in the year 1986 indicate that habitats were more fragmented. According to Forman and Godron (1986), MPS affects biomass, primary productivity, nutrient storage per unit area, as well as species composition and diversity. Variability in patch size (i.e. MPSCV) measures a key aspect of landscape heterogeneity that is not captured by MPS. The patches in the year 1986 are much more variable in size than those in other years, suggesting increased landscape heterogeneity. Greater variability indicates less uniformity, and may reflect differences in underlying processes affecting the landscapes.

Landscape fragmentation expressed by NP and MPS, can also be revealed through the connectance index (CONNECT). The physical continuity of habitats across the landscape (structural connectedness), represented by the connectance index (CONNECT), decreased from 0.47% in 1957 to 0.17% in 2009 within a 200 m threshold distance. The low percentage value of connectance could arise from the dominance of agriculture over time or almost none of the patches in the landscape are "connected" within a 200 m threshold distance of another patch of the same type.

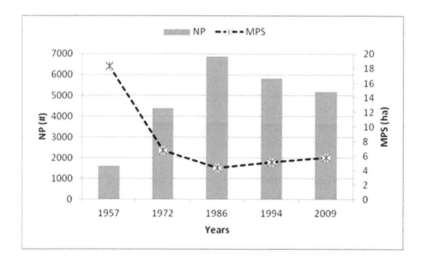

Figure 16 The relative values of NP and MPS over time explaining higher fragmentation in 1986

At landscape level the area-weighted mean shape index (AWMSI) quantifies landscape configuration in terms of the complexity of patch shapes. The AWMSI indicates that the patches in the 1986 landscape are most irregular in shape, whereas the patches in 1957 landscape are least irregular in shape. The mean shape-weighted values for all years considered are greater than the unweighted values (i.e. AWMSI>MSI), indicating that larger patches in each landscape are more irregular in shape than the average. The primary ecological consequence of shape complexity is its effect on plant species richness. Torras (2008) reported negative correlation between AWMSI and plant species richness. Patch shape complexity, expressed by AWMSI, increased from 10.1 in 1957 to 20.65 in 2009 suggesting a decrease in plant species richness over the landscape. As shape complexity increases, the entire patch mosaic becomes increasingly disaggregated (i.e., less contagious).

4.5.2 Changes in landscape pattern at class level

Habitats such as riverine forest, wetlands, natural forest and rangelands are particularly important for the ecological functioning of the ecosystem. The decline in the habitat extent and fragmentation of these critical habitats might bring quite different dynamics of some ecological processes. Therefore, the spatial pattern analysis of these critical habitats will help the understanding of the human activities that might cause the loss of these habitats. Table 15 shows the landscape pattern of ecologically important habitats of the Jedeb watershed.

Riverine forest: Both percentage of land (PLAND) and mean patch size (MPS) of riverine forest have decreased from 4% to 1.08% and 3.58 ha to 1.29 ha, respectively, during the period 1957-2009 (Table 15). The gradual decreases in both PLAND and MPS along with the increase in number of patches until 1986 suggests that this habitat was highly subdivided into smaller riverine forest fragments especially in 1986. Consequently, the riverine landscape could not support more species. Clumpiness index (CLUMPY) which offers a useful index of aggregation independent of patch type abundance revealed that the patches of the 1957 riverine forest were more clumped in distribution than random. The lower CLUMPY value in 1986 (CLUMPY=0.67) indicates that riverine landscapes under the 1986 conditions were more disaggregated (subdivided) than expected under a spatially random distribution. The 2009 riverine forest patches are more isolated than the 1957 patches as the mean nearest-neighbour distance (ENN_MN) has increased from 147 m in 1957 to 253 m in 2009. The inter-patch connectivity has also decreased. This pattern is mirrored by a decrease in mean proximity index (PROX_MN) from 12.67 in 1957 to 3.64 in 2009. The above results indicate that the riparian vegetation has undergone considerable fragmentation during the study period.

Riparian vegetation mediates the exchange of water, nutrients, sediments and energy between the terrestrial and aquatic ecosystems. The degree to which riparian forest provides its functions is related to the mean patch size and inter-patch connectivity of riparian forest (Shirley and Smith, 2005). The loss and fragmentation of riparian forest affects its landscape functions. Wide and unfragmented riparian forests have a greater role in hydrologic regulation and maintaining water quality. Rriparian forest traps and filters sediments and nutrients from runoff that contains agro-chemicals coming from adjacent cropland before they reach the stream. This filtration role of riparian forest is dependent on patch size and connectivity of the riparian forest (Binford and Buchenau 1993; Ward and Stanford 1995).

Marshland: The percentage of land (PLAND) of the marshland decreased from 3.86% in 1957 to 2.29% in 2009 indicating a decrease in habitat extent over time (Table 15). The splitting index (SPLIT) supports this conclusion as it increases from 4768 in 1957 to 81,636 in 2009 indicating that marshlands have increasingly been subdivided into smaller patches. Marshlands were highly fragmented in 1986 and 1994 as indicated by higher values of the number of patches (NP=927, 867) and a smaller mean patch size (MPS=0.85ha, 0.84ha). Marshland habitat loss and fragmentation can further be mirrored by the increase in disaggregation of patches over time as indicated by the decrease in clumpiness index (CLUMPY) from 0.73 in 1957 to 0.63 in 1986 and 0.62 in 1994.
Most of the marshlands in the Jedeb watershed are located along the Jedeb river and its tributaries. These marshlands are now being converted to cropland in order to meet the increasing demand for food as a result of population growth. Land redistribution in the Amhara region has not carried out since 1996. However, casual conversations with some of the local farmers revealed that there is informal land redistribution by local government bodies to the landless youth in the watershed. This informal land redistribution is focused on marshlands that are located close to rivers for the purpose of irrigation. Wetland loss and fragmentation resulting from socio-economical and institutional drivers will eventually impair some ecological functions of these wetlands.

Natural woody vegetation: Shrubs and bushes, woodland and *ericaceous* forest together would form natural woody vegetation. PLAND of natural forest (SHB, WL and EF) decreased from 11.5% in 1957 to 2.8% in 2009, suggesting that these rare and critical landscape features need to be considered in the maintenance of biodiversity in the landscape.

Rangeland: Rangeland comprises grassland and afro-alpine grassland. NP of grassland drastically increased from 169 in 1957 to 2007 in 1986 and slightly decreased to 1835 in 2009. MPS after a sudden drop from 35.97 ha in 1957 to 4.57 ha in 1972, grassland showed a gradual decrease to 2.4 ha in 2009 (Table 15). CLUMPY value decreased from 0.88 in 1957 to 0.69 in 1986 and slightly increased to 0.71 in 2009 suggesting that the patches of grassland have become increasingly disaggregated and less clumped from 1957 onwards. The structural connectedness of grassland decreased as indicated by the decrease in CONNECT values from 1.18 in 1957 to 0.14 in 2009. The mean proximity index (PROX_MN) mirrored the patterns of CONNECT regarding the inter-patch connectivity. Mean shape index (MSI) decreased from 1.86 in 1957 to 1.31 in 1994 and slightly increased to 1.33 in 2009, indicating that the current patches of grassland are more regular in shape than those of the past. This result indicates fragmentation of grassland caused a simplification of patch shapes. The landscape patterns of afro-alpine grassland and grassland are more or less similar except patch size variation. A large difference in values between MPS and AWMPS of afro-alpine grassland indicates that patches are on average quite large, although there are many small patches. The watershed contains a wide diversity of sizes of afro-alpine grassland patches as discerned by a large variation in patch size distribution (MPSCV). Therefore, it is important for conservation planners to be aware of the presence of large patches along with small and medium sized patches.

Cultivated land: Percentage of landscape (PLAND) indicates that in all years the Jedeb landscape comprises more than 50% of cultivated land (Table 15). Therefore, cultivated land constituted a matrix because it is extensive and highly connected, and likely to exert a dominant influence on the flora and fauna and ecological processes. NP value increased from 91 in 1957 to 624 in 1986 and decreased to 220 in 2009. Cultivated land was highly fragmented in 1986 as indicated by a higher value of the number of patches (NP=624) and smaller mean patch size (MPS=31.55 ha). The lower value of clumpiness (CLUMPY=0.76) also shows the 1986 patches of cultivated land are increasingly disaggregated and less clumped because of the strengthened fragmentation process. A large difference in values between MPS and AWMPS indicates patches of cultivated land are on average quite large, although there are many small patches. The watershed contains a wide diversity of sizes and shapes of cultivated land patches as discerned by a large variation in patch size distribution (MPSCV).

Table 15 Class level landscape metrics

TYPE	YEAR	PLAND	NP	ED	MPS	AWMPS	MPSCV	MSI	AWMSI	PROX_MN	ENN_MN	CLUMPY	CONNECT	SPLIT
RF	1957	4	331	13.26	3.58	21.15	221	1.55	2.66	12.67	147	0.75	0.51	35047
	1972	2.06	457	8.71	1.34	9.13	241	1.27	1.96	2.65	201	0.69	0.26	157451
	1986	2	496	8.72	1.19	14.82	338	1.24	2.42	4.55	177	0.67	0.31	100235
	1994	1.77	372	7.08	1.41	15.1	311	1.24	2.35	3.86	208	0.7	0.34	110722
	2009	1.08	248	4.78	1.29	6.63	203	1.28	2	3.64	253	0.68	0.51	414085
ML	1957	3.86	496	13.61	2.31	22.54	296	1.33	2.42	4.8	145	0.73	0.28	34108
	1972	3.09	255	9.63	3.6	21.81	225	1.47	2.22	3.59	247	0.77	0.39	43944
	1986	2.67	927	13.18	0.85	4.6	209	1.18	1.7	2.82	118	0.63	0.22	241591
	1994	2.47	867	12.32	0.84	4.6	211	1.19	1.71	2.45	118	0.62	0.24	260527
	2009	2.29	682	10.7	1	5.02	201	1.21	1.74	2.98	132	0.65	0.27	257691
EF	1957	0.94	8	0.94	35.01	156.2	186	1.53	2.94	215.21	249	0.94	28.57	20103
	1972	0.68	11	1	18.38	130.65	247	1.65	3.59	65.44	374	0.9	10.91	33318
	1986	0.44	11	0.6	11.92	83.31	245	1.43	2.28	52.42	156	0.92	10.91	80544
	1994	0.21	7	0.27	8.9	51.3	218	1.16	1.76	7.96	217	0.94	23.81	274795
	2009	0.2	7	0.27	8.55	48.05	215	1.26	1.73	8.67	117	0.94	28.57	306253
WL	1957	3.93	179	9.52	6.52	45.46	244	1.65	2.56	12.03	198	0.82	0.66	16584
	1972	2.89	554	10.84	1.55	13.2	274	1.24	1.95	2.25	210	0.72	0.21	77708
	1986	2.63	938	12.91	0.83	7.69	287	1.18	1.82	1.99	175	0.63	0.15	146452
	1994	1.36	314	5.54	1.28	9.21	249	1.23	1.91	1.51	285	0.7	0.24	236706
	2009	1.27	379	6.03	1	4.7	193	1.22	1.69	1.14	275	0.65	0.23	496200
SHB	1957	6.65	318	17.24	6.2	41.21	238	1.66	2.82	13.44	161	0.79	0.38	10816
	1972	3.15	1112	16.36	0.84	4.22	201	1.2	1.69	1.37	184	0.6	0.11	223369
	1986	2.37	972	12.2	0.72	6.88	292	1.16	1.73	1.67	174	0.61	0.15	181920
	1994	1.97	650	9.37	0.9	8.52	291	1.2	1.89	2.11	201	0.64	0.17	176279
	2009	1.33	280	5.09	1.41	10.94	260	1.24	2.02	2.61	307	0.72	0.31	204691
GL	1957	20.5	169	26.47	35.97	638.53	409	1.86	6.53	541.3	150	0.88	1.18	227
	1972	19.91	1292	58.2	4.57	84.89	419	1.46	4.43	38.28	120	0.73	0.21	1755
	1986	16.62	2007	58.24	2.46	42.61	404	1.35	3.49	29.71	103	0.69	0.15	4188
	1994	14.24	1899	48.79	2.22	29.1	348	1.31	2.72	14.59	113	0.7	0.14	7151
	2009	14.86	1835	49.84	2.4	51.35	451	1.33	3.4	21.95	113	0.71	0.14	3889
AGL	1957	6.19	11	2.93	166.78	1807.32	314	1.99	6.04	707.14	100	0.96	29.09	265
	1972	5.91	106	8.66	16.53	496.91	539	1.49	6.71	208.89	120	0.88	2.34	1010
	1986	5.27	166	7.27	9.42	1050.22	1051	1.3	8.31	456.23	105	0.89	1.64	536
	1994	5.21	223	6.75	6.92	1046.93	1226	1.24	6.97	345.6	108	0.89	1.22	543
	2009	4.87	163	6.05	8.87	810.7	951	1.3	5.45	263.93	154	0.9	1.47	751
CL	1957	53.41	91	36.94	174.04	9974.69	750	1.86	14.73	10600.78	114	0.89	2.05	5.57
	1972	61.78	442	65.78	41.46	17097.54	2028	1.26	32.83	13066.92	114	0.79	0.61	2.81
	1986	66.36	624	70.73	31.55	18161.75	2397	1.24	33.27	25217.06	86	0.76	0.51	2.46
	1994	69.28	296	62.82	69.35	19535.5	1675	1.3	30.62	27171.06	88	0.78	0.96	2.19
	2009	69.47	220	55.98	93.7	20016	1458	1.31	28.31	25930.82	100	0.8	1.15	2.13

RF = Riverine Forest, ML = Marsh Land, EF = Ericaceous Forest, WL = Woodland, SHB = Shrubs and Bushes, GL = Grassland, AGL = Afro-alpine Grassland, CL = Cultivated Land

4.6 CONCLUSIONS

The landscape pattern analysis revealed that the Jedeb landscape as a whole showed a considerable change in landscape composition and configuration over the period 1957-2009. Cultivated land constituted a matrix (i.e. $> 50\%$ of landscape) and likely to exert a dominant influence on the flora, fauna and many ecological processes of the watershed. This increase in dominance over time has caused several classes to occur in a clumped distribution and decreased spatial diversity of the watershed. More habitat loss and fragmentation observed in the 1986 landscape was associated with more irregularity in patch shapes (higher spatial heterogeneity). However, fragmentation of rangeland and marshland caused simplifications of patch shapes. Therefore, the interpretation that fragmentation has led to complexity of patch shapes is not a ubiquitous relationship. The observed changes in composition and configuration of riverine forest, marshland, natural forest and rangeland could influence some major landscape functions such as plant species diversity, hydrologic regulation and water quality soil erosion and water infiltration. Close monitoring of land use by land planners is required for prompt mitigation of such adverse effects of landscape functions as a result of landscape pattern change. To conclude, when it is not possible or feasible to conduct field surveys, application of landscape metrics based on basic ecological principles could contribute to landscape planning and monitoring. Future studies should focus on the relationship between the identified landscape metrics and different ecological processes based on detailed field survey data.

Chapter 5

SATELLITE-BASED MONITORING OF WETLAND

CHANGES AND THEIR ECOHYDROLOGICAL

CHARACTERISATION IN THE MT.CHOKE RANGE[3]

5.1 ABSTRACT

Wetlands provide multiple ecosystem services such as storing and regulating water flows and water quality, providing unique habitats to flora and fauna, and regulating micro-climatic conditions. Conversion of wetlands for agricultural use is a widespread practice in Ethiopia, particularly in the south-western part where wetlands cover large areas. Although there are many studies on land cover and land use changes in this region, comprehensive studies on wetlands are still missing. Hence, extent and rate of wetland loss at regional scales is unknown. The objective of this paper is to quantify wetland dynamics and estimate wetland loss in the Choke Mountain range (area covering 17,443 km^2) in the Upper Blue Nile basin, a key headwater region of the river Nile. Therefore, satellite remote sensing imagery of the period 1986-2005 was considered. To create images of surface reflectance that are radiometrically consistent, a combination of cross-calibration and atmospheric correction (Vogelman-DOS3) methods was used. A hybrid supervised/unsupervised classification approach was used to classify the images. Overall accuracies of 94.1% and 93.5% and Kappa Coefficients of 0.908 and 0.913 for the 1986 and 2005 imageries, respectively, were obtained. The results showed that 607 km^2 of seasonal wetland with low moisture and 22.4 km^2 of open water areas are lost in the study area during the period 1986 to 2005. The current situation in the wetlands of Choke Mountain is characterized by further degradation, which calls for wetland conservation and rehabilitation efforts through incorporating wetlands into watershed management plans.

---

[3] *This chapter is based on* Teferi, E., Uhlenbrook, S., Bewket, W., Wenninger, J., Simane, B., 2010. The use of remote sensing to quantify wetland loss in the Choke Mountain range, Upper Blue Nile basin, Ethiopia. *Hydrol. Earth Syst. Sci.* 14, 2415–2428.

5.2 INTRODUCTION

It is well documented that wetlands play critical roles in the hydrological system (Ehrenfeld, 2000; Tiner, 2003). Moderating the impact of extreme rainfall events and providing base flow during dry periods is one of the key functions of wetlands. On the one hand, wetlands can moderate flow dynamics and mitigate flooding. On the other hand, they play a regulating role on the water quality by capturing sediment and capturing and converting pollutants. Wetland functions are particularly important in areas with high rainfall variability as they help to sustain smaller discharges during the dry season and, consequently, improve the availability of water. Therefore, wetland conservation has a vital role to play in alleviating water problems at different scales (Jensen et al., 1993; Reimold, 1994).

Estimates of global wetland area range from 5.3 to 12.8 million km^2 (Finlayson et al., 1999; Matthews and Fung, 1987). About half the global wetland area has been lost as a result of human activities (OECD, 1996). In tropical and subtropical areas conversions of wetlands to alternative land uses have accelerated wetland loss since the 1950s (Moser et al., 1996) and agriculture is considered the principal cause for wetland loss. The degradation and loss of headwater wetlands leads to increased flooding, soil erosion, degradation of water quality, reduced dry season flows, lower groundwater tables and less availability of water and moisture during the dry season. In Ethiopia due to limited research work on wetland resources, the management of wetlands is impaired by the general public and decision makers. Hence many of the wetlands are drying up and disappearing; a good example is the collapse of Lake Alemaya in Eastern Ethiopia (Lemma, 2004).

In the Blue Nile basin, wetlands in general are given limited attention and the role of headwater wetlands is especially not well addressed. This seems to be somewhat different for the White Nile basin, where the ecosystem functioning and its services were studied in the Lake Victoria region (Loiselle et al., 2006; Van Dam et al., 2007) and the Sudd wetland in Sudan (Mohamed et al., 2005). It is common that small headwater wetlands are ignored in wetland studies and often the importance of large wetlands is recognized and appreciated. But these seemingly insignificant individual wetlands can collectively play important roles in moderating flows and improving water quality (McKergow et al., 2007). A large number of such small wetlands, ranging from sedge swamps to seasonally flooded grasslands, exist in the Mt. Choke headwater areas. However, very little is known about the spatial distribution, the variability in space and time and their hydrological and ecological functioning. Most previous research on wetland was conducted in south-west Ethiopia with a focus on wetland management and policy implications (Hailu et al., 2003; Wood, 1996), hydrological impacts of wetland cultivation (Dixon, 2002; Dixon and Wood, 2003), socio-economic determinants (Mulugeta et al., 2000), and gender dimension of wetland use (Wood, 2003).

An effective and efficient management of wetlands requires an exhaustive survey (mapping) of their distribution and determination of whether or not they have changed over time and to what extent (Baker et al., 2006; Jensen et al., 1993). Ground-based survey of wetlands of large, or even small, wetlands is very time consuming. The use of remote sensing techniques offers a cost effective and time saving alternative for delineating wetlands over a large area compared to conventional field mapping methods (Ozesmi and Bauer, 2002; Töyrä and Pietroniro, 2005). Landsat, Satellite Pour l'Observation de la Terre (SPOT), Advanced Very High Resolution Radiometer (AVHRR), Indian Remote Sensing satellites (IRS), radar systems (Ozesmi and Bauer, 2002), Advanced Space-borne Thermal Emission and Reflection Radiometer (ASTER) (Pantaleoni et al., 2009) and Moderate-resolution Imaging Spectro-radiometer (MODIS) (Ordoyne and Friedl, 2008) are the most frequently used satellite sensors for wetland detection. However, there is no standard method for computer-based wetland classification (Frazier and Page, 2000).

The aim of this study is to quantify wetland dynamics and estimate wetland loss in the Choke Mountain range. To achieve this objective, a hybrid supervised/unsupervised classification of Landsat imagery acquired in 1986 and 2005 were considered. Before classification it was necessary to create images of surface reflectance that are radiometrically consistent and to ensure inter-image comparability between TM and ETM+ images. This was done by applying a combination of cross-calibration and atmospheric correction (Vogelman-DOS3) methods (Paolini et al., 2006).

5.3   DEFINITION OF WETLANDS

Successful identification of wetlands starts with a clear definition of wetlands. Numerous wetland definitions have been developed for various purposes. Most recent ones are definition given by RAMSAR and USGS. The Ramsar convention (Ramsar Convenon Secretariat, 2006) defined wetlands as: "....*areas of marsh, fen, peat land or water, whether natural or artificial, permanent or temporary, with water that is static or flowing, fresh, brackish or salt, including areas of marine water the depth of which at low tide does not exceed six metres.*" The United States Geological Survey (USGS) defined wetland as a general term applied to land areas which are seasonally or permanently waterlogged, including lakes, rivers, estuaries, and freshwater marshes; an area of low-lying land submerged or inundated periodically by fresh or saline water (Mac et al., 1998). Taking the RAMSAR and USGS definitions into consideration, the wetlands mapped in this study included water bodies (static or flowing water), seasonal wetlands with high moisture, and seasonal wetlands with low moisture. Riparian vegetation was not included in the wetland classes of this study.

5.4 MATERIALS AND METHODS

5.4.1 The Study Area

The study area covers the whole of the Choke Mountain range, the most important source of water for the Blue Nile river system in Ethiopia. The area extends between 10° to 11°N and 37°30' to 38°30'E, the highest peak is located at 10°42'N and 37°50'E. It is situated in the south of Lake Tana, in the central part of the Amhara National State of Ethiopia (Fig. 17). Elevation extends from 810 m a.s.l to 4050 m a.s.l. The mean annual rainfall varies between 995 mm a^{-1} and 1864 mm a^{-1} based on data from 13 stations for the years 1971-2006. The mean annual temperature ranges from 7.5°C to 28°C (BCEOM, 1998d). The Choke Mountain is the water tower of the region serving as headwater of the upper Blue Nile basin. Many of the tributaries of the Upper Blue Nile originate from this mountain range. A total of 59 rivers, and many springs are identified in the upper catchments of Choke Mountain.

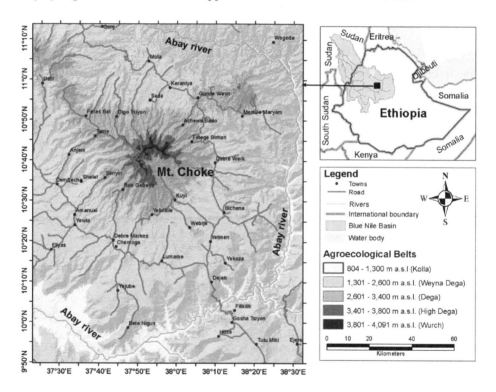

Figure 17 Location map of Mt. Coke range in the Upper Blue Nile basin

The main geological unit in the area is the Tarmaber Gussa formation which represents Oligocene to Miocene basaltic shield volcanoes with minor trachyte and phonolite intrusions (Tefera et al., 1999). The soil units covering the Choke Mountain are Haplic Alisols (deep soils with predominant clay or silt-clay texture), Eutric Leptosols (shallow soils with loam or clay-loam texture) and Eutric Vertisols (deep soils with clayey texture and angular/sub-angular blocky structure) (BCEOM, 1998b). Due to active morphological processes (erosion, landslides etc.) the soil depth can vary between zero and several meters.

There is no longer significant natural forest cover in this mountain range. The major remaining natural habitats are moisture moorland, sparsely covered with giant lobelias (Lobelia spp.; Jibara/Jibbra), lady's mantle (*Alchemilla spp.*), Guassa grass (*Festuca spp.*) and other grasses. There is very little natural woody plant cover; heather (*Erica spp.*; Asta) and Hypericum (*Hypericum revolutum*; Amijja) are found in patches. Bamboo or Kerkeha (*Arundinaria alpina*) is found as homestead plantation as well as part of the natural vegetation cover in the area, albeit very sparsely. Korch (*Erithrina brucei*) is commonly grown as border demarcation plant in the area. *Eucalyptus globulus is* extensively grown in plantation, and some of the residents have become dependent on it for their livelihoods.

5.4.2   Data Used and Image Pre-processing

The study utilized 27 topographic maps of the area at a scale of 1:50,000 and 1:250,000 dated 1984, geological maps at the scale of 1:250,000, and Landsat images of years 1985/6 (TM) and GLS 2005 (ETM+). Wet and dry season images of each year were acquired. Scenes were required to be of the same phenological cycle (dry or wet season) and have little or no cloud cover. The characteristics of the image data are shown in Table 16. All Landsat images were accessed free of charge from U.S. Geological Survey (USGS) Centre for Earth Resources Observation and Science (EROS) via http://glovis.usgs.gov/. All scenes supplied by the EROS Data Centre had already been georeferenced to the Universal Transverse Mercator (UTM) map projection (Zone 37), WGS 84 datum and ellipsoid. The GLS 2005 Landsat 7 ETM+ image was processed to Standard Terrain Correction (Level 1T -precision and terrain correction) in the USGS -EROS using the Level 1 Product Generation System (LPGS). The Landsat 5 TM (1986) standard product is processed using the National Land Archive Production System (NLAPS). Besides, sixteen-day Moderate Resolution Imaging Spectroradiometer (MODIS) derived Normalized Difference Vegetation Index (NDVI) at a resolution of 250 m gained from USGS, Land Processes Distributed Active Archive Centre, to characterize the final wetland classes.

Several pre-processing methods were implemented before the classification and the change detection. These include geometric correction, radiometric correction, atmospheric correction, topographic normalization and temporal normalization. The importance of each pre-processing steps has been discussed extensively in the literature (Han et al., 2007; Hill and Sturm, 1991; Vicente-Serrano et al., 2008). The specific pre-processing steps followed in this research are briefly described below.

Table 16 Description of Landsat images used

	1986		2005	
	Dry Season	**Wet Season**	**Dry Season**	**Wet Season**
Collection type	Landsat C.	Landsat C.	GLS collection	Landsat SLC-on
Satellite	Landsat 5	Landsat 5	Landsat 7	Landsat 7
Sensor	TM	TM	ETM+	ETM+
Path/Row	169/053	169/053	169/053	169/053
Acquisition date	28/01/86	9/11/1985	24/11/05	15/10/02
Pixel spacing (m)	28.5	28.5	30	30
Sun Elevation	44.52	50.425	50.7467	58.87
Sun Azimuth	128.79	132.73	141.585	125.86

Geometric correction: Re-projection to the local level projection system was made (UTM, map projection; Clarke 1880, Spheroid; and Adindan Datum). In some parts of the study area, there were significant discrepancies between the imageries and the underlying GIS base layers, which were extracted from high resolution SPOT-5 imagery. The misaligned scenes were georectified to the underlying base layers by using control points.

In this study, first the GLS 2005 Landsat 7 ETM+ was geometrically rectified using 25 control points taken from the topographic map at 1: 50,000 and 11 GPS points taken in the field. The other images were then registered to the ETM+ image with the same projection. An RMS error of less than 0.5 was achieved. The nearest neighbour resampling method was used to avoid altering the original pixel values of the image data. Because of the high incidence of varied topography in the study area, ortho-correction was found to be necessary and hence undertaken to further enhance the image geometry by accounting for the significant spatial distortion caused by relief displacement.

Radiometric Calibration: The absolute radiometric correction method used in this paper involved a combination of cross-calibration method developed by Vogelmann et al. (2001), with an atmospheric correction algorithm based on COST model (Chavez, 1996), denoted DOS3 by Song et al. (2001). Paolini et al. (2006) called this combination of methods "Vogelman-DOS3".

Cross-calibration: When using both Landsat TM and ETM+ images in studies that require radiometric consistency between images, special attention has to be paid to the differences in sensors response. Previous studies have demonstrated significant differences in the radiometric response between Landsat ETM+ and Landsat TM spectral bands (Teillet et al., 2001; Vogelmann et al., 2001). To reduce these differences and ensure image inter-comparability, a cross-calibration was needed (Paolini et al., 2006; Vicente-Serrano et al., 2008). In the cross-calibration procedure, the Landsat TM images DN (DN5) were first converted to the Landsat ETM+ DN (DN7) images using the equation developed by Vogelmann et al. (2001). After this conversion, the L5 TM DN is treated as L7 ETM+ DN, and L7 ETM+ gain, offset and ESUN values are used respectively for radiance and Top-Of-Atmosphere (TOA) reflectance.

Conversion to at-sensor spectral radiance: At-sensor spectral radiance was computed for the cross-calibrated Landsat 5-TM (1986 images) quantized calibrated pixel values in DNs and Landsat 7-ETM+ quantized calibrated pixel values in DNs using sensor calibration parameters published by Chander et al. (2009) and in image header file.

Conversion to Top of Atmosphere (TOA) reflectance: To correct for illumination variations (sun angle and Earth-Sun distance) within and between scenes, TOA reflectance for each band was calculated. This conversion is a very important step in the calculation of an accurate Normalized Difference Vegetation Index (NDVI) and other vegetation indices. Some of the parameters for the conversion are available in the image header files, while the exo-atmospheric irradiance values for Landsat 7 are published by Chander et al. (2009).

Atmospheric correction
TOA reflectance value does not take into account the signal attenuation by the atmosphere, which strongly affects the inter-comparability of the satellite images taken on different dates. But atmospheric correction methods account for one or more of the distorting effects of the atmosphere and thereby convert the brightness values of each pixel to actual reflectance as they would have been measured on the ground.

COST atmospheric correction model is one of the models have been developed to eliminate atmospheric effects to retrieve correct physical parameters of the earth's surface (e.g. Surface reflectance) (Chavez, 1996). Despite the variety of available techniques, a fully image-based technique developed by Chavez (1996) known as the COST model that derives its input parameters from the image itself, has been proved to provide a reasonable atmospheric correction (Berberoglu and Akin, 2009). The COST atmospheric model was used to convert the at-sensor spectral radiance to reflectance at the surface of the Earth.

Topographic correction
In mountainous regions such as Upper Blue Nile Basin, topographic effects that result from the differences in illumination due to the angle of the sun and the angle of terrain produce different reflectance for the same cover type. Hence, the classification process for rugged terrain is seriously affected by topography and needs careful treatment of the data (Hale and Rock, 2003; Hodgson and Shelley, 1994).

To reduce the topographic effects methods such as spectral-band Ratioing (Holben and Justice, 1981), application of a Lambertian model (Cosine Correction) (Smith et al., 1980) and application of a non-Lambertian (Minnaert Correction and C-correction) (Meyer et al., 1993) have been developed. The C-correction (Teillet et al., 1982) is chosen for this study to correct for uneven reflectance patterns, as it is the most effective method of reducing the topographic effects for Landsat imagery (Meyer et al., 1993). The cosine of the incident solar angle was calculated using the equation of Holben and Justice (1981)and Smith et al. (1980). The variables used in the equation were derived from a digital elevation model (DEM), and sun orientation information was gathered from image header data. The Regress module in IDRISI calculated the C-correction coefficient by regressing actual radiance values of each image band (the dependent variable) and cosine of the incident solar angle (the independent variable). The derived C-correction coefficients (Table 17) were then used as parameters to normalize both images for topographic effects. Besides visual analysis, the evaluation of topographic correction was based on the computation of R^2-values for the linear regression between corrected data and cosine of the incident solar angle. Since the topographic impact varies with wavelength and is most prevalent in the near-infrared spectral region, a reduction in the R^2 value for the near-infrared band was observed (0.214 to 0.008 for the 1986 image and 0.291 to 0.0342 for the 2005 image). This indicates the effectiveness of the non-Lambertian C-correction.

Table 17 C-correction coefficients derived for the non-Lambertian corrected images

Year of image	Band 1	Band 2	Band 3	Band 4	Band 5	Band 7
1986	0.397	0.326	0.49	0.261	0.597	0.498
2005	0.341	0.42	0.486	0.327	0.63	0.584

Temporal Normalisation

Vicente-Serrano et al. (2008) recommended an additional step (temporal normalization) to completely remove non-surface noise and improve temporal homogeneity of satellite imagery. Therefore 9 reservoir sites and 5 bare sites were used to normalize the 2005 ETM+ image (slave) with respect to 1986 TM image (reference). The resulting normalization equation is shown in Table 18.

Table 18 Image normalization regression models developed for Choke area image.

	Regression models	r^2	Normalization targets
1986 TM B2	=-25.2 + 1.03 (2005 ETM+ B2)	0.981	8 wet and 4 dry
1986 TM B3	=-26.4 + 1.03 (2005 ETM+ B3)	0.993	9 wet and 5 dry
1986 TM B4	=-21.7 + 0.98 (2005 ETM+ B4)	0.987	8 wet and 5 dry

Image Classification

A hybrid supervised/unsupervised classification approach was used to classify the images. This approach involved: (1) unsupervised classification using ISODATA (Iterative Self-Organizing Data Analysis) to determine the spectral classes into which the image resolved; (2) using the spectral clusters ground truth (reference data) were collected to associate the spectral classes with the cover types observed at the ground for the 2005 image and for the 1986 image reference data were collected from the 1984 Topographic map (1:50,000 scale); and (3) classification of the entire image using the maximum likelihood algorithm. A hierarchical class grouping was adopted to label and identify the classes (Thenkabail et al., 2005) in both the wet and dry season images. First, the unsupervised ISODATA clustering was used. 25 initial classes were found and then after a rigorous identification and labelling process the original 25 classes were progressively reduced to 17, 7 and 4 classes by using ground truth data. Ground truth data on spatial location, land cover types, soil moisture status, and topographic characteristics were collected from selected sample sites during the period from June 2009 to March 2010. Stratified random sampling design was adopted for the selection of 342 point sample sites. Stratification was based on the accessibility of the sites. Spatial locations were obtained using a GPS (Global Positioning System).

To assist within class identification process in the unsupervised classes, the at-sensor reflectance based NDVI (Normalized Difference Vegetation Index) values were calculated. Wetlands with covers of barren lands and/or with sparse vegetation have lower NDVI values as a result of soil moisture that is relatively higher than the surrounding areas. When wetlands have vegetation or crops the NDVI will vary depending on vegetation density and vigour. At-sensor reflectance based Tasseled Cap wetness values (Huang et al., 2002) were also calculated for highlighting wetland/ non-wetland areas from the spectral classes Tasseled Cap coefficients of 0.2626 (B1), 0.2141 (B2), 0.0926 (B3), 0.0656 (B4), -0.7629 (B5), and -0.5388 (B7) were used.

Post-classification Processing and Assessment
A variety of change detection techniques have been developed (Lu et al., 2004;Yuan and Elvidge, 1998). Each of them has its own merit and no single approach is optimal and applicable to all cases. Yuan and Elvidge (1998) divide the methods for change detection and classification into pre-classification and post-classification techniques. In this paper post-classification comparison change detection approach which compares two independently produced classified land use/cover maps from images of two different dates was used(Jensen, 2005). The principal advantage of post-classification is that the two dates of imagery are separately classified; hence it does not require data normalization between two dates (Singh, 1989). The other advantage of post-classification is to provide information about the nature of change (including trajectories of change) (Song et al., 2001).

The accuracy of the classification results was assessed by computing the confusion matrix (error matrix) which compares the classification result with ground truth information. Seventeen classes were identified using the hybrid supervised/unsupervised classification. Classes of different spectral patterns but the same information classes were selectively combined in the classified images and finally reduced to 7 classes. Majority analysis was also performed to change spurious pixels within a large single class to that class. Finally, a detailed tabulation of changes in

wetland and non-wetland classes between 1986 and 2005 classification images were
compiled by post-classification comparison

Table 19 Descriptions of the land cover classes identified in the Mt. Choke range

Class Code	Class Name	Description
OW	Open water	Ponds or stagnant water
SWL	Seasonal wetlands with low moisture (Islam *et al.*, 2008)	wetlands, grasslands covered with less
SWH	Seasonal wetlands with high moisture (Islam *et al.*, 2008)	wetlands, grass lands and very moist farmlands or irrigated land (high vegetation cover and high NDVI)
AF	Afro alpine herbaceous vegetation	Afro alpine grasses
WV	Woody vegetation	shrubs, bushes, plantations
CL	Cultivated land	rainfed agriculture
BL	Bare land	exposed soil and rock

5.5 RESULTS AND DISCUSSION

The study area was classified into seven classes: three wetland classes and four non-wetland classes. The description for each class is given in Table 19. Although riparian vegetation cover defines one type of wetland, because of the limitation of the classification method used it was classified as woody vegetation.

A Confusion Matrix was computed to assess the accuracy of the classification by comparing the classification results with ground truth information. From the Confusion Matrix report 94.1% and 93.5% overall accuracy for the 1986 and 2005 imageries were attained, respectively. The overall accuracy was calculated by summing the number of pixels classified correctly and dividing by the total number of pixels. An overall accuracy level of 85% was adopted as representing the cutoff point between acceptable and unacceptable results according to Anderson et al. (1976). The Confusion Matrix also reports the Kappa Coefficient (k) which is another measure of the accuracy of the classification (Cohen, 1960). The k values for the 1986 and 2005 image classification were 0.908 and 0.913, respectively. The values can range from -1 to $+1$. However, since there should be a positive correlation between the remotely sensed classification and the reference data, positive k values indicate the goodness of the classification. Landis and Koch (1977) characterized the possible ranges for k into three groupings: a value greater than 0.80 (i.e., 80%) represents strong agreement; a value between 0.40 and 0.80 (i.e., 40–80%) represents moderate agreement; and a value below 0.40 (i.e., 40%) represents poor agreement.

The accuracies for both the 1986 and 2005 image classification were considered good enough to apply post-classification change detection analysis. Basically this analysis focuses on the initial state classification changes (1986 image). For each initial state class (1986), the post-classification comparison technique identifies the classes into which those pixels changed in the final state image (2005).

5.5.1 Space-time Dynamics of Non-wetlands

Table 20 lists the initial state classes in the columns, the final state classes in the rows and the unchanged areas in the diagonal cells. For each initial state class (i.e., each column), the table indicates how these pixels were classified in the final state image. A total of 891 km^2 was classified as woody vegetation in the initial state image (1986), while 2965 km^2 was classified as woody vegetation in the final state image (2005) (Table 20). 2074 km^2 woody vegetation cover has increased over the 20 years studied. This is mainly attributed to the increase in *Eucalyptus* plantation in the area (e.g. Raba Forest and Wildlife Reserve). After significant deforestation in the past the scarcity of wood for fuel and other uses has forced the local people to plant trees (Bewket, 2003). About 46% (1351 km^2) of the existing woody vegetation has come from cultivated land and 30 % (873.2 km^2) from bare land categories. Since the productive capacity of the the cultivated land is being threatened by the loss of nutrients through erosion, farmers in the area are looking for another option because they could not cope with declining crop yields. Every year more and more agricultural land is being converted into *Eucalyptus* forest plantation which ensures income security for local residents. This trend in land-use and land-cover change is altering the soil (Bewket and Stroosnijder, 2003) and hydrologic (Bewket, 2005) characteristics of upland watersheds of the Choke Mountain range. This may also influence the livelihoods of the population living in downstream areas by changing critical watershed functions (e.g availability of water during the dry season).

Table 20 Change detection statistics

		Initial State Image (1986)							
		OW	SWH	SWL	AF	WV	CL	BL	Class Total
Initial State Image (1986)	OW	15.4	25	0.5	0	0	0.1	5.4	46.4
	SWH	27.3	748.7	275.3	52.2	122.9	1621.7	151.5	2999.6
	SWL	0	132.3	111.5	15.1	10.9	64	5.7	339.5
	AF	0	96	13.7	79.9	49.9	47.3	12.4	299.2
	WV	17	301.4	59.3	34.1	328.7	1351	873.2	2964.7
	CL	6.9	632.8	422	38.3	307.2	6687.3	674.3	8768.8
	BL	2.2	104.7	64	7.6	71.2	1002.7	772.1	2024.5
	Class Total	68.8	2040.9	946.3	227.2	890.8	10774.1	2494.6	

OW=open water, SWH=seasonal wetlands with high moisture, SWL=seasonal wetlands with low moisture, AF=afro alpine herbaceous vegetation, WV=woody vegetation, CL=cultivated land, BL=bare land

76

5.5.2 Space-time Dynamics of Wetlands

In the initial state (1986) image, 946.3 km² of land was classified as seasonal wetlands with low moisture, but only 339.5 km² of land was classified as seasonal wetlands with low moisture in the final state image (Fig. 18 & 19). This means a total of 607 km² of seasonal wetlands with low moisture were lost and/or converted to another land cover/use type during the period 1986 to 2005 (Table 20). Out of the 946.3 km² of seasonal wetlands with low moisture class in the initial state image, only 111.5 km² (12%) remain unchanged and 835 km² (88 %) of the initial class has been converted and this makes the class most dynamic (unstable) of all observed land cover types. About 422 km² (~51%), 275.3 km² (~33%), 64 km² (8%), and 59.3 km² (7%) of the previous seasonal wetlands with low moisture cover were converted to cultivated land, seasonal wetland with high moisture, bare land, and woody vegetation respectively. Every year 21 km² of seasonal wetlands with low moisture land cover types are converted to cultivated land. This indicates that the experience of wetland cultivation which is common in south-wet Ethiopia (Dixon, 2002; Dixon and Wood, 2003) is also practiced in Choke Mountain range. With the increase in population pressure and shortage of land, government considers wetland cultivation as a remedy to address the food needs (Wood, 1996). As a result, seasonal wetlands with low moisture have come under pressure.

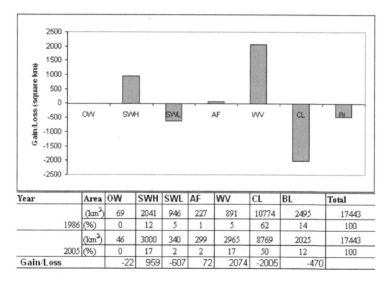

Year		Area	OW	SWH	SWL	AF	WV	CL	BL	Total
		(km²)	69	2041	946	227	891	10774	2495	17443
1986	(%)	0	12	5	1	5	62	14	100
		(km²)	46	3000	340	299	2965	8769	2025	17443
2005	(%)	0	17	2	2	17	50	12	100
Gain/Loss			-22	959	-607	72	2074	-2005	-470	

Figure 18 Proportions of land cover types in 1986 and 2005. The definition of the abbreviations of land cover types can be found at Table 19.

Again based on Table 20, 2040.9 km² of land was classified as seasonal wetlands with high moisture in the initial state image, and 2999.6 km² of land was classified as seasonal wetlands with high moisture in the final state image. A total gain of 959 km² in the seasonal wetlands with high moisture land cover was observed (Fig. 18). Out of

2999.6 km^2 seasonal wetland with high moisture class, 54% (1621.7 km^2) have come from cultivated land, 9% (275.3 km^2) have come from seasonal wetlands with low moisture class and 25 % (748.7 km^2) remain unchanged. The contribution of cultivated land to the increase in the seasonal wetland with high moisture class is significant (54%). Since the gain is mostly from classes initially classified as cultivated land, the increase in moist farmlands (irrigated land) caused the increase in the seasonal wetland with high moisture class. This is related to the expansion of small scale irrigation (man-made wetland) in the 1990s. Jedeb irrigation project located at Yewla village (East Gojam) constructed in 1996 by diverting Jedeb river through a diversion weir (11° 22' N and 37° 33' E) is one of the best examples.

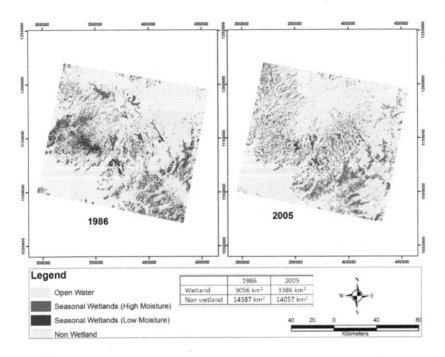

Figure 19 Spatial distribution of wetlands

In the initial state (1986) image, 68.8 km^2 of land was classified as open water and 46.4 km^2 of land was classified as open water in the final state image. This means a total of 22.4 km^2 of open water wetlands were lost and converted to another land cover/use type during the period 1986 to 2005. About 40 % (27.3 km^2) of the initial open water class were converted to seasonal wetlands with high moisture, 25% (17 km^2) to bare land and 22% (15.4 km^2) remain unchanged. According to farmers of the locality Wezem pond and Ginchira pond located in Muga watershed of Choke Mountain are now being converted to bare land. This was also substantiated by field observation. In contrast, Bahirdar pond (Fig. 20) at Yenebirna village in Chemoga watershed showed a gradual increase through time (Table 21). To analyse this situation 10 satellite images (5 on dry season and 5 on wet season) were taken and the size of Bahirdar pond was determined. The observed smallest size (4.2 ha) could be because of the 1986 drought occurred throughout Ethiopia. This indicated that the condition of wetlands of the area is dependent on climatic factors beyond human influences. The increasing trend of rainfall of the area around the pond (Fig. 21) associated with increasing runoff and sediment inflow from the surrounding watershed could be one of the reasons for the gradual increase of the pond. A study by McHugh et al. (2007) explained one example from north-east Ethiopia for the case of Hara swamp.

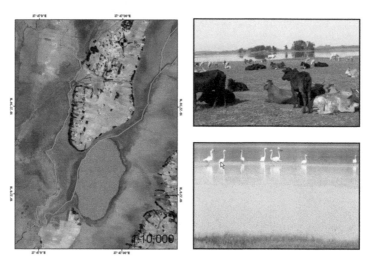

Figure 20 Open water wetland at Yenebrna village (East Gojam)

Table 21 Time-series of pond size

	1984	1986	1995	2000	2005/6
Dry season	7 ha	4.2 ha	23.1 ha	34 ha	35 ha
Wet season	7.4 ha	6.2 ha	39 ha	39 ha	41 ha
Source	Landsat TM	Landsat TM	Landsat TM	Landsat ETM+	ASTER

5.5.3 Temporal Characterisation of Wetlands

Although Landsat (TM and ETM+) has 16-day repeat cycle, these 16-day images are not available for the period 2001-2009. Therefore, the temporal characteristics of wetlands and non-wetlands were studied using Moderate Resolution Imaging Spectroradiometer (MODIS), 16-day Normalized Difference Vegetation Index (NDVI)

79

at a resolution of 250 m gained from United States Geological Survey, Land Processes Distributed Active Archive Centre.

Open water cover type shows lower NDVI values than the other wetland classes throughout the year. Because open water features have lower near infrared reflectance. In open water class higher NDVI values are observed during the period of June – August (rainy season). The increase in runoff with increase in trapped sediment during rainy season may increase the NDVI response. Most cover types cannot be distinguished during the rainy season. Seasonal wetlands with high moisture and low moisture classes are clearly distinct during the period from January to May. Thus, wetland mapping can be effective at a time of year when the different wetland types can be best distinguished. The NDVI time-series class characteristics follows the rainfall trend (Fig. 21); vegetation begins to gain vigour in beginning of September and remains at high vigour until November while the rainfall reaches maximum during 28 July to 12 August. Previous studies have investigated the lag and cumulative effect of rainfall on vegetation in semi-arid and arid regions of Africa (Martiny et al., 2006; Wagenseil and Samimi, 2006) and humid regions of Africa (Onema and Taigbenu, 2009).

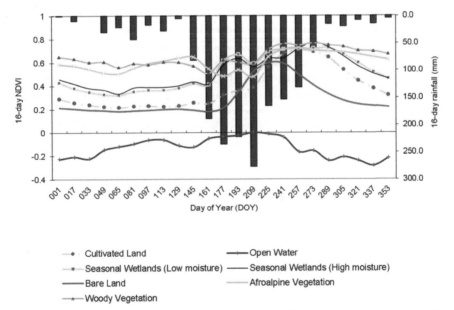

Figure 21 The time-series characteristics of the wetlands and non-wetlands (Mean 16-day annual NDVI and rainfall for the period 2001 to 2009)

5.6 CONCLUSIONS

The study demonstrated the use of remote sensing techniques to delineate headwater wetlands from non-wetlands and determine the dynamics over large areas of Choke Mountains with the overall accuracy of 94.1% and 93.5% and Kappa Coefficient of 0.908 and 0.913 for the 1986 and 2005 imageries, respectively. Therefore, ground-based wetland surveying for large area, especially for small wetlands is a very time consuming and ineffective process. The use of Remote Sensing and GIS techniques in wetland mapping reduces cost and enhances the accuracy. Creating radiometrically consistent image data through (Vogelman-DOS3) methods before applying hybrid supervised/unsupervised classification was found to be effective in wetland mapping. However, the application of the technique to Landsat images was not effective in mapping riparian vegetation as wetland. To undertake a more focused wetland management plan implementation, the use of mandatory environmental characteristics (vegetation, soil, and hydrology) for wetland detection is still important.

Four major trajectories of change that were observed: (1) from seasonal wetlands with low moisture to cultivated land, (2) from cultivated land to woody vegetation (plantation), (3) from open water to bare land, and (4) from bare land to cultivated land. In general, 607 km^2 of seasonal wetland with low moisture and 22.4 km^2 of open water are lost in the study area over the 20 years considered. This is one indication of future deterioration in wetland condition. This calls for wetland conservation and rehabilitation through incorporating wetlands into watershed management plans so as to make wetlands continue providing their multiple functions which include flood control, stream-flow moderation, groundwater recharge, sediment detention, and pollutant retention. Further research on specific wetland spots is needed for accurate inventories of wetlands and to explore the reasons why and how wetlands are changing.

PART 2: CHANGES IN SOIL HYDROLOGY IN

RESPONSE TO LULCC

Figure 22 Field visit to the research site

Chapter 6

EFFECTS OF LAND USE AND LAND COVER ON

SELECTED SOIL QUALITY INDICTORS IN THE JEDEB

WATERSHED [4]

6.1 ABSTRACT

Soil degradation is a major environmental and economic problem in Ethiopia. Soil degradation follows changes in land use and land cover (LULC) that generally involves expansion of agricultural land uses into steep lands or areas previously considered as marginal lands. Understanding changes in soil quality resulting from land use and land management changes is important for designing sustainable land management plans or related interventions. This study evaluated the influence of land use and land cover (LULC) on key soil quality indicators (SQIs) within a small watershed (Jedeb, 296.6 km^2) in the Blue Nile Basin of Ethiopia. Factor analysis based on Principal Component Analysis (PCA) was used to determine different SQIs. Surface (0-15 cm) soil samples with four replications were collected from five main LULC types in the watershed (i.e. natural woody vegetation, plantation forest, grassland, cultivated land, and barren land) and at two elevation classes (upland and midland), and thirteen soil properties were measured for each replicate. A factorial (2x5) multivariate analysis of variance (MANOVA) showed that LULC and altitude together significantly affected OM levels. However, LULC alone significantly affected

---

[4] *This chapter is based on* Teferi, E., Bewket, Simane, B., W., Uhlenbrook, S., Wenninger, J. (2015). Effects of Land Use and Land Cover on Selected Soil Quality Indicators (Environmental Monitoring and Assessment: under review).

bulk density and altitude alone significantly affected bulk density, soil acidity, and silt content. Afforestation of barren land with *Eucalyptus* trees can significantly increase the soil OM in the midland part but not in the upland part. Soils under grassland had a significantly higher bulk density than did soils under natural woody vegetation indicating that de-vegetation and conversion to grassland could lead to soil compaction. Thus, the historical LULC change in the Jedeb watershed has resulted in the loss of soil OM and increased soil compaction. The study shows that a land use and management practices can be monitored if it degrades, maintains or improves the soil using key soil quality indicators.

6.2 INTRODUCTION

Land degradation has become an important global concern because of its implications for food security and the environment (Nabhan, 2001). One of the major causes of land degradation is human activities. Human activities contributing to land degradation include deforestation, removal of natural vegetation, overgrazing, and agricultural practice without erosion control measures. Land use and land cover (LULC) change is the easiest detectable indicator of human activities on the land. LULC change affects soil quality (Biro et al., 2013; Yu et al., 2012), runoff and sedimentation rates (Leh et al., 2013; Liu et al., 2012b), biodiversity (Hansen et al., 2004b), and ecosystem services (Zhang et al., 2013).

Soil degradation is the key component of land degradation and there is almost no form of land degradation that does not include soil degradation (Hartemink, 2003). It is the decline in soil quality leading to a reduction in other components of land resources (e.g. vegetation, water, air). Soil degradation processes involve: chemical degradation (e.g. nutrient depletion and acidification); physical degradation (e.g. soil erosion, compaction and waterlogging); and biological degradation (e.g. soil organic matter decline and depletion of soil fauna) (Lal et al., 1989). Regarding soil nutrient depletion, Ethiopia is among the sub-Saharan African (SSA) countries with very high nutrient depletion rates. For example, the nutrient balances for Ethiopia were: -41 kg/ha in 1983 and -47 kg/ha in 2000 for N; -6 kg/ha in 1983 and -7 kg/ha in 2000 for P; and -26 kg/ha in 1983 and -32 kg/ha in 2000 for K (Misra et al., 2003). This very high nutrient depletion is an indication of the prevailing unsustainable land use and management systems. In Ethiopia, land is used largely by smallholders for subsistence agriculture, which is characterized by low yields resulting from little or no use of external inputs. In the absence of intensification, smallholders always seek more land for cultivation to meet the increasing demand for food caused by the growing population. Consequently, deforestation and conversion of grassland to cropland takes place including in marginal lands. This leads to high soil erosion rates.

In the Ethiopian highlands (e.g. East Gojjam), soil degradation is severe (Bewket and Teferi, 2009; Hurni, 1988). Farmers are responding to soil degradation by changing land use from cropland to plantation forest (e.g. *Eucalyptus*). Apart from the importance of plantation forest for fuel-wood, its long-term effects on the soil have received little research attention. Little systematic effort has been undertaken to prove whether or not plantation forest is a more sustainable form of land use than crop cultivation(e.g.Bewket and Stroosnijder, 2003). A land use and management system should be evaluated, if it degrades, maintains or improves soil quality so as to support the design of site-specific and effective interventions that are aimed at enhancing farmer's ability to address nutrient deficits. Interventions need to be based on purpose- and site-specific soil quality indicators (Shukla et al., 2006; van der Zaag, 2009). Soil quality indicators used elsewhere may not be applicable to another area.

Soil quality has two different aspects: (i) intrinsic soil quality which refers to the soil's natural composition, and (ii) dynamic soil quality covering the part influenced by soil users or managers (Larson and Pierce, 1994). In other words, the dynamic soil quality is influenced by changes in land use and land cover or land management practices. For instance, removal of vegetative covers exposes the soil to the forces of erosion, increase soil temperatures, increases compaction, decreases sources of soil organic matter and so on. These effects will ultimately lead to negative changes in soil quality indicators from the perspective of sustainable land use. Understanding changes in soil quality resulting from land use and land management changes is important as it provides information on the effectiveness of different land use options and hence modifies land management practices as needed to maintain or improve soil quality for sustainable land use.

The general objective of this study was to assess effects of land use and land management on selected key soil quality indicators. The specific objectives were to: (i) identify suitable key soil quality indicators using multivariate analysis, and (ii) assess effects of land use and land management changes on the selected soil quality indicators taking the influence of agroecology into account.

6.3 MATERIALS AND METHODS

6.3.1 Study Area

The study area, Jedeb watershed, lies between 10°23' to 10°40' N and 37°33' to 37°60' E. It is situated in the south-western part of Mount Choke, which is an important headwater area of the Upper Blue Nile/Abay in Ethiopia (Fig. 23). Administratively, the area belongs to two *weredas* (districts; *Machakel* and *Sinan*) of East *Gojam* Zone in *Amhara* Regional State of Ethiopia. Land and livestock are the basic sources of livelihood for the farmers in the Jedeb watershed, who are engaged in small scale and subsistence mixed agriculture. The watershed (about 296 km^2 in size) comprises diverse topographic conditions with elevation ranging from 2172 m to nearly 4001 m

(Fig. 23), and slopes ranging from nearly flat ($<2°$) to very steep ($>45°$). Altitude is an important parameter in soil genesis, as it provides a useful proxy for several factors such as geomorphology and climate of places.

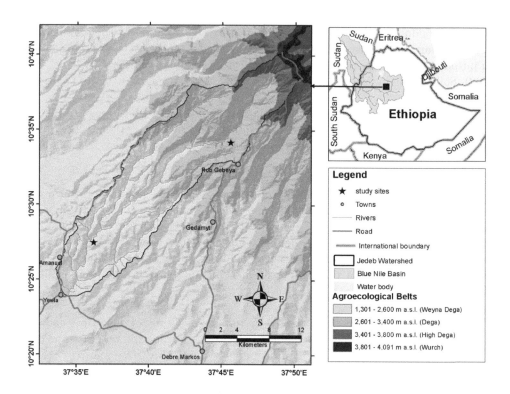

Figure 23 Location of the study are and agroecological belts of the site

The soil units covering the Jedeb watershed are Haplic Alisols (54%), Eutric Leptosols (24%), and Haplic Luvisols (22%) (BCEOM, 1998b). The geological units in the watershed are Tarmaber Basalts-2 (less weathered, 94.2% of the watershed), Tarmaber Basalts-1 (deeply weathered and appropriate for agricultural use, 4.4%), Blue Nile Basalt (0.9%), and Alluvium (0.5%) (BCEOM, 1998e).

6.3.2   Land use and land cover change history

In this study, five land use and land cover (LULC) types were used for the purpose of assessing effects of land use and land cover on soil properties (Table 22). The LULC classes were obtained by aggregating the ten LULC classes in the watershed identified by Teferi et al. (2013). Black and white aerial photographs of 1957, Landsat image 2009 (TM) were used to derive the ten LULC classes, and to evaluate the changes between 1957 and 2009. Details of land use and land cover change analysis of the Jedeb watershed are given in Teferi et al. (2013). A considerable land use and land cover change was observed in the Jedeb watershed during the period 1957-2009. Cultivated land constituted the predominant type of land cover and it accounted for approximately 53% and 70% in the years 1957 and 2009, respectively. Cultivated land showed a ~27% gain of the total area, while natural woody vegetation and grassland showed losses of ~13% and ~20%, respectively (Fig. 24). Plantation forest also increased from zero in 1957 to 3.4% of the watershed in 2009. The contribution of cultivated land to the increase in plantation forest was 1.45% of the total (largest proportion) (Teferi et al., 2013). This indicates that local people plant *Eucalyptus* trees on their croplands instead of growing crops. The local people took this option when they believed that their land had become less fertile and not economical for crop production.

Table 22 The five major land use and land cover types studied in the Jedeb watershed

Aggregated land use and land cover classes*	1957		1986		2009	
	km^2	%	km^2	%	km^2	%
Natural Woody Vegetation (NWV)	46.1	15.5	22.1	7.5	11.5	3.9
Plantation Forest (PF)	0	0	2.1	0.7	10	3.4
Grassland (GL)	90.6	30.6	72.7	24.5	65.3	22
Cultivated land (CL)	158.4	53.4	197.2	66.5	206.2	69.5
Barren land (BL)	1.5	0.5	2.6	0.9	3.6	1.2

*The original 10 land use and land cover classes of Teferi et al. (2013) were aggregated into 5 classes. Woodland, Ericaceous Forest, Riverine Forest, and Shrubs and bushes were grouped into NWV; Grassland, Afro-alpine grassland, Marshland classes were aggregated into GL; and Plantation forest, Cultivated land, and Barren land classes remain unchanged.

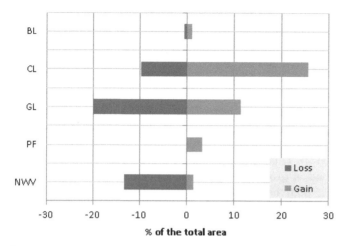

Figure 24 Gains and losses of land use and land cover classes between 1957 and 2009 based on Teferi et al. (2013). NWV (natural woody vegetation); PF (plantation forest); GL (grassland); CL (cultivated land); BL (barren land).

6.3.3 Soil Sampling and Analysis

Soil sampling

Five main LULC types and two elevation classes were selected to study effects of land use and land cover on key soil quality indicators. The LULC types were natural woody vegetation, plantation forest, grassland, cultivated land, and barren land (Table 22). The elevations of the specific sampling sites range from 2400–2500 m (midland) and 2900–3000 m (upland), which respectively represent the Dega and WeynaDega agroecological zones (AEZs) (Fig. 23). For the unimodal rainfall region of Ethiopia, which also includes the Jedeb watershed, Hurni (1999) identifies the following agroecological zones: WeynaDega (mid-altitude, 1300-2600 m a.s.l.), Dega (high-altitude, 2600-3400 m a.s.l.), High Dega (high-altitude, 3400-3800 m a.s.l.), and Wurch (>3800 m a.s.l.). The selected sites have a land use history of at least 23 years. The crop type on the cultivated land in both AEZs was *engido* (*Avena spp.*). The selected type of forest plantation was *Eucalyptus globulus* in both the midland and upland areas, as it is a widely planted tree in the Jedeb watershed.

Site selection was conducted based on 1:250,000 scale Soil Mapping Units (SMUs) developed by (BCEOM, 1998b) because each mapping unit is characterized by similar major landform and major soils. One SMU (V/SeLp) that forms similar landform and soil type across the two elevation classes was selected from a total of 3 SMUs identified in the Jedeb watershed. The geology of all sampling sites was the Miocene (~27 to 5 My) Tarmaber Basalts-2 formation (BCEOM, 1998e). Soil samples were taken from similar soil types (i.e. Leptosls) and slope classes (i.e. 15-30%) at the two elevation classes.

In a factorial completely randomized design considering the five land use types and the two elevation classes with four replications, a total of 40 different sampling units (n = 5x2x4 = 40) were created. For each replicate, a sample was collected for laboratory analysis within a 15 m x 15 m plot in the layer 0-15 cm (topsoil) using soil auger. Undisturbed soil samples were taken using cylindrical metal samplers for the purpose of soil bulk density determination using core method. Figure 25 shows how he filed sampling was undertaken. The collected disturbed soil samples were air-dried, lightly ground and passed through a 2 mm sieve, and subsequently analysed in laboratory for texture, pH, Organic Carbon (OC), Total Nitrogen (TN), exchangeable bases (sodium, potassium, calcium, and magnesium), Cation Exchange Capacity (Dinku et al.), and available phosphorus (Av.P). Particle size analysis (percentage of sand, silt and clay) was determined by the Hydrometer method (Bouyoucos, 1962). Soil pH was measured using a 1:5 soil: water suspension method (Rayment and Higginson, 1992). Available Phosphorus (Av.P) was extracted according to the Bray 2 method (Bray and Kurtz, 1945). Total Nitrogen was determined by the Kjeldahl method (Bremner and Mulvaney, 1982). Soil OC was determined by the Wakley and Black rapid titration method (Nelson and Sommers, 1982). Exchangeable bases (Ca^{2+}, Mg^{2+}, K^+, and Na^+) were determined using the Ammonium Acetate method (Thomas, 1982). CEC was determined using a method based on Chapman (1965) which is recommended distinctly for acid soils.

Figure 25 Soil sampling using soil augering (a) and soil profile (b) in the Jedeb watershed

Infiltration measurements

Field infiltration measurements were carried out using double-ring infiltrometer in each land use and land cover types with three replicates so that the measurement is realistic. The rings were driven 5 cm into the soil. The area inside the cylinder was flooded with water obtained from the nearby river and the rate at which the water infiltrates into the soil was measured until the intake rate had become nearly constant.

In this study, infiltration data obtained from the field experiments were fitted to infiltration model of Philip (Philip, 1957). The model-estimated final infiltration rate was computed to be used in ANOVA. The two-parameter Philip equation is given by

$$i(t) = \frac{1}{2}St^{-0.5} + B \qquad\qquad [5.1]$$

where: $i(t)$ is the infiltration rate at time t (cm hr^{-1}), B is transmissivity (cm hr^{-1}), and S is the sorptivity (cm hr$^{-0.5}$).

Statistical analysis

Various steps were followed to identify key soil quality indicators (SQIs). Principal component analysis (PCA) was used as the method of factor extraction to group the measured 10 soil properties into major varimax rotated components of the SQI. The first principal component (PC) accounts for as much of the variability in the data as possible, and each succeeding component accounts for as much of the remaining variability as possible. Only factors with eigenvalues >1 were retained based on Kaiser's recommendation (Kaiser, 1960). Factor scores for each sample were computed using the regression method and analysis of variance was conducted on the computed factors to determine, which factors varied significantly with land use and land cover types. A two-way between-subjects multivariate analysis of variance (MANOVA) was conducted on the soil properties with high loadings (i.e. BD, pH, OM, and silt) in the statistically significant factors. One of the major advantages of using MNAOVA is that it can single out group differences that may become masked with univariate statistical analyses.

6.4 RESULTS AND DISCUSSION

6.4.1 Grouping of Soil Properties

A principal component analysis (PCA) was conducted on the 10 soil parameters with orthogonal rotation (varimax). The Kaiser–Meyer–Olkin (KMO) measure verified the sampling adequacy for the analysis, KMO = 0.71, which is above the acceptable limit of 0.5 (Field, 2009). Bartlett's test sphericity was highly significant (χ^2 (45) = 164. 45, p <0.001), indicating that the correlations of the soil parameters were sufficiently large for PCA (Table 23).

Table 23 Pearson correlation coefficients of the physical and chemical properties

	Silt	Clay	BD	pH	CEC	BS	OM	TN	C:N
Silt%	1								
Clay%	-.777***	1							
BD (g cm⁻³)	-.412***	.625***	1						
pH (-)	-.479***	.465***	.464***	1					
CEC (cmol kg⁻¹)	-.068	-.109	-.092	.176	1				
BS (%)	-.164	.139	-.025	-.102	.192	1			
OM (%)	.663***	-.757***	-.640***	-.540***	.207*	-.060	1		
TN (%)	.263*	-.366***	-.319**	-.028	.190	-.061	.235*	1	
C:N (-)	.263*	-.157	.015	-.229*	-.057	.000	.215*	-.596***	1
Av.P (ppm)	-.019	-.170	-.265**	.119	.423***	.100	.233*	.089	.010

The determinant of the correlation matric is 0.009.
* correlation is significant at 0.1, ** correlation is significant at 0.05, *** correlation is significant at 0.01 level of probability

Four components with eigenvalues greater than the Kaiser's criterion of 1 were retained and in combination explained 78.82% of the variability in the measured soil properties (Table 24). Component-1 explained nearly 36% of variance with a high loadings (>0.7) from silt content, clay content, OM, BD and pH (Fig. 26). Silt content was highly significantly correlated with Clay (r = -0.78), OM (r = 0.66), BD (r = -0.41), and pH (r = -0.48). Component-2 explained 18% of the total variance with high loadings from C:N and TN, resulting from the highly significant correlation between them (r = -0.596). Component-3 accounted for nearly 15% of the total variance with high loadings from available P and CEC, resulting from the highly significant negative correlation between them (r = 0.423). Component-4 explained 10% of the total variance with high positive loading from base saturation.

The amount of variance in each soil parameter that can be explained by the retained 4 components is represented by the communality estimates. As can be seen in the communality estimates of Table 24, the four components explained >86% of variance in BS, C:N and TN; >82% in organic matter and clay content; >70% in soil pH, available P and silt content; and >65% in bulk density and CEC.

A univariate ANOVA conducted on the component scores revealed that there were statistically significant main effects of LULC and altitude and interaction effect on component-1 only (S1). The other components were not significant, implying that the variation of the soil properties with high loadings in these components was not related to LULC or altitude. Therefore, only soil properties with high loadings in component-1 were analysed applying MANOVA.

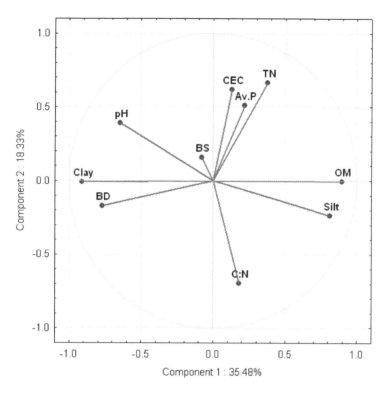

Figure 26 Loading plot showing relationships among soil properties in Component-1 and Component-2, Jedeb watershed, central highlands of Ethiopia

Table 24 The factor loadings after Varimax rotation and communality estimates (Factor loadings over 0.7 appear bold)

Soil parameters	Rotated component loadings				Communality estimates
	1*	2	3	4	
Clay (%)	**.894**	.073	-.127	.169	.848
OM (%)	**-.877**	.050	.223	-.023	.822
Silt (%)	**-.833**	.055	-.108	-.225	.759
BD (g cm⁻³)	**.749**	.193	-.174	-.150	.651
pH (-)	**.718**	-.141	.338	-.340	.765
C:N (-)	-.231	**.910**	.015	-.043	.883
TN (%)	-.314	**-.863**	.111	-.086	.863
Av.P (ppm)	-.114	.026	**.832**	.014	.706
CEC (cmol kg⁻¹)	-.020	-.102	**.815**	.123	.690
BS (%)	.067	.020	.135	**.933**	.893
Initial Eigenvalues	3.548	1.833	1.458	1.043	-
Variance (%)	35.485	18.328	14.578	10.428	-
Cumulative variance (%)	35.485	53.812	68.390	78.817	-

* Effects of land use and land cover (LULC), altitude (A) and interaction of the two (LULC x A) on component-1 are significant ($p < 0.05$).

6.4.2   Results of Statistical Assumptions Tests

The results of the evaluation of the assumptions of normality, linearity, and multicollinearity were satisfied. Based on a statistically nonsignificant Box's test (Box's $M = 83.01$, $P = 0.073$) (S2), the assumption of homogeneity of variance-covariance matrices was met. This means that the dependent variables (BD, pH, OM, Silt) covariance matrices were equal across the levels of the independent variables (LULC and altitude). Separate Levene's tests (S2) for each dependent variable found statistically non-significant ($P > 0.05$) effect for all dependent variables, indicating equal variances for each dependent variable across the levels of LULC and altitude. Bartlett's test of sphericity was statistically significant ($\chi^2 = 425$, $P = 0.000$), indicating a sufficient correlation between the dependent variables to proceed with the MANOVA analysis.

6.4.3   Multivariate Main Effect and Interaction Effect

Table 25 presents the multivariate test results, all of which indicate statistically significant main effects (LULC, altitude) as well as the interaction effect. Since equality of variance-covariance matrices was evidenced with the non-significant Box's M test, the most robust and commonly used statistic, the Wilks' Lambda (Λ), was selected for further analysis. Using Wilks's criterion, the multivariate main effect of land use was statistically significant ($\Lambda = 0.23$, $F (16, 83) = 3.206$, $P = 0.000$), indicating the soil properties were significantly affected by land use. The multivariate main effect of altitude was also statistically significant ($\Lambda = 0.235$, $F (4, 27) = 21.98$, $P = 0.000$). The multivariate interaction effect of LULC and altitude was statistically significant ($\Lambda = 0.388$, $F (16, 83) = 2.311$, $P = 0.033$).

Table 25 Multivariate test results

Effect	Multivariate Test	Value	F	Hypothesis df	Error df	Sig.
Land use and	Pillai's Trace	1.015	2.549	16	120	.002
land cover	Wilks' Lambda	0.230	3.206	16	83	.000
(LULC)	Hotelling's Trace	2.366	3.770	16	102	.000
	Roy's Largest Root	1.924	14.434	4	30	.000
Elevation (E)	Pillai's Trace	0.765	21.984	4	27	.000
	Wilks' Lambda	0.235	21.984	4	27	.000
	Hotelling's Trace	3.257	21.984	4	27	.000
	Roy's Largest Root	3.257	21.984	4	27	.000
LCLC * E	Pillai's Trace	0.806	1.892	16	120	.027
	Wilks' Lambda	0.388	1.888	16	83	.033
	Hotelling's Trace	1.129	1.800	16	102	.041
	Roy's Largest Root	0.553	4.144	4	30	.009

A series of univariate ANOVAs were conducted on each dependent variable separately as a follow-up test to the MANOVA in order to determine the locus of the statistically significant multivariate effects. Since the Leven's test (S2) showed the homogeneity of variances among the groups on each dependent measure, it was possible to proceed with univariate ANOVAs with a Bonferroni adjusted alpha level of 0.025 (0.05/2). As can be seen in Table 26, the univariate interaction effect of altitude and land use on soil organic matter content was statistically significant ($F = 3.884$, $p = 0.012$).

Although the mean of the interaction of elevation and land use differs significantly on soil organic matter content, there is no indication of the location of this difference. To test for differences among specific interaction levels, Least Significant Difference (LSD) post hoc comparisons were conducted on a new grouping variable that contains the ten levels of interaction. Out of 45 possible combinations of interactions only 23 of them were found to be significantly different from each other with respect to soil organic matter content (Table 27).

Table 26 Univariate test results

Source	Dependent Variable	F	Sig.
Land use and land cover (LULC)	Bulk density	5.994	.001
	Soil pH	2.376	.074
	Organic matter	11.067	.000
	Silt content	0.969	.439
Elevation (E)	Bulk density	40.796	.000
	Soil pH	24.403	.000
	Organic matter	28.587	.000
	Silt content	22.814	.000
LULC * E	Bulk density	2.428	.070
	Soil pH	0.818	.524
	Organic matter	3.884	.012
	Silt content	1.509	.225

A statistically significant mean difference in soil organic matter content between soils under midland natural woody vegetation and midland barren land was amongst the highest (mean = 4.55%). There was a statistically significant mean difference in soil organic matter content between soils under upland cultivated land and midland cultivated land (Mean Difference = 2.7, p = 0.000) (Fig. 27). Soils under upland cultivated land significantly contained more soil organic matter (mean = 5.26%) than under midland cultivated land (mean = 2.56%). The results also indicate that soils under upland natural woody vegetation had significantly higher soil organic matter content than those under all midland land use types except midland natural woody vegetation.

Since no other soil properties were statistically significant in the univariate interaction effect, it is possible to proceed to examine only the main effects of bulk density, soil pH and silt content. The main effect of land use was statistically significant only for bulk density ($F = 5.994$, $p = 0.001$). The LSD post hoc test (Table 27) suggested that grassland (mean = 1.13 g cm^{-3}) and barren land (mean = 1.20 g cm^{-3}) had a significantly higher bulk density than did soils under natural woody vegetation (mean = 0.94 g cm^{-3}). Barren land had a significantly higher bulk density as compared to the remaining land cover types.

The main effect of altitude was statistically significant for soil pH ($F = 24.403$, $p = 0.000$), bulk density ($F = 40.796$, $p = 0.000$), and silt content ($F = 6.613$, $p = 0.015$). The midland part had a significantly higher soil pH (mean = 5) and bulk density (mean = 1.19 g cm^{-3}) than the upland part (mean pH of 4.5 and mean bulk density of 0.95 g cm^{-3}) of the watershed. The silt content of the upland part (mean = 40.5%) was significantly higher than that of the midland part (mean = 26.9%). The upland has a clay loam soil texture and the midland has clay soil texture. The mean physical and chemical soil properties of the upland and midland of the Jedeb watershed are shown in S3.

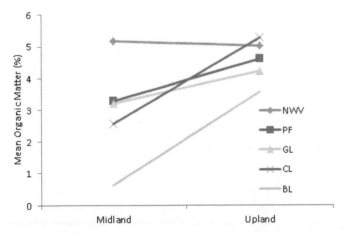

Figure 27 Interaction plot for the mean soil organic matter (OM) content. NWV (natural woody vegetation); PF (plantation forest); GL (grassland); CL (cultivated land); BL (barren land)

6.4.4 Effects of land use and land cover on infiltration rate

The infiltration data was Log transformed to make it normally distributed. A one-way ANOVA of the log-transformed data indicates that there is a statistically significant difference ($F = 17.45$, $p < 0.001$) between the infiltration rates on the four different land uses. Figure 28 shows the median value of the average infiltration rate for different land use and land cover types. The soils under natural woody vegetation have relatively higher infiltration rates (3.14 cm hr^{-1}), while cropland and grassland show relatively lower infiltration rates. Further analysis utilized the Least Significant Difference (LSD) test to distinguish significantly different mean infiltration rates between land use categories. Statistically significant differences in infiltration rates were evident between the following land use and land cover types: plantation forest and grassland ($p < 0.01$), plantation forest and natural woody vegetation ($p < 0.05$), plantation forest and cultivated land ($p < 0.01$), cultivated land and natural woody vegetation ($p < 0.001$), and grassland and natural woody vegetation ($p < 0.001$).

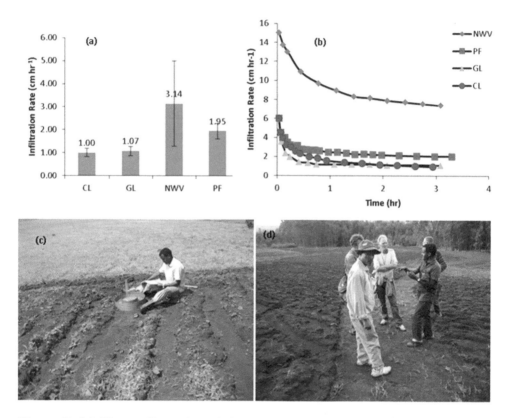

Figure 28 (a) The median value of the average infiltration rate (cumulative intake divided by total time) for different land use and land cover, (b) instantaneous infiltration rate, (c) a photo taken during field infiltration test, and (d) a photo showing the compacted soil layer that hinders infiltration after the removal of topsoil due to erosion on a cropland of Mt. Choke area. NWV (natural woody vegetation); PF (plantation forest); GL (grassland); CL (cultivated land); BL (barren land)

Table 27 Multiple comparison post hoc LSD test for the interaction effect of land use and altitude with respect to soil organic matter and for the main effect of land use on bulk density

Soil parameters	(I) Interaction	(J) Interaction	Mean Difference (Zhang et al.)	Sig.
Soil organic matter content	Upland natural woody vegetation	Upland barren land	1.465	.041
	Upland natural woody vegetation	Midland plantation forest	1.725	.008
	Upland natural woody vegetation	Midland grassland	1.803	.013
	Upland natural woody vegetation	Midland cultivated land	2.451	.001
	Upland natural woody vegetation	Midland barren land	4.388	.000
	Upland plantation forest	Midland cultivated land	2.041	.009
	Upland plantation forest	Midland barren land	3.978	.000
	Upland grassland	Midland cultivated land	1.655	.017
	Upland grassland	Midland barren land	3.591	.000
	Upland cultivated land	Upland barren land	1.714	.014
	Upland cultivated land	Midland plantation forest	1.975	.002
	Upland cultivated land	Midland grassland	2.053	.004
	Upland cultivated land	Midland cultivated land	2.700	.000
	Upland cultivated land	Midland barren land	4.637	.000
	Upland barren land	Midland barren land	2.923	.000
	Midland natural woody vegetation	Upland barren land	1.624	.024
	Midland natural woody vegetation	Midland plantation forest	1.884	.004
	Midland natural woody vegetation	Midland grassland	1.962	.008
	Midland natural woody vegetation	Midland cultivated land	2.610	.001
	Midland natural woody vegetation	Midland barren land	4.547	.000
	Midland plantation forest	Midland barren land	2.663	.000
	Midland grassland	Midland barren land	2.585	.000
	Midland cultivated land	Midland barren land	1.937	.006
Bulk density	Grassland	Natural woody vegetation	0.144	.022
	Barren land	Natural woody vegetation	0.284	.000
	Barren land	Plantation forest	0.178	.006
	Barren land	Grassland	0.140	.026
	Barren land	Cultivated land	0.166	.009

6.5   DISCUSSION

Analysis of variance on the principal components revealed that soil texture, organic matter content, bulk density and soil pH were found to be appropriate soil quality indicators that are related to land use or/and altitude. The non-significant effects on the other components suggested that the variation in soil properties of the corresponding components could not be explained either by land use or altitude. The findings of this study are consistent with that of Tesfahunegn (2013), who investigated soil organic carbon, silt content and bulk density as important parameters for the evaluation of sustainability of land use. Shukla et al. (2006) also found a similar result that soil organic carbon could play an important role for monitoring soil quality in relation to land use.

6.5.1   Effects of Land Use and Land Cover on Soil Quality Indicators

Soil organic matter content

The impact of LULC on soil organic matter content was dependent on altitude, as indicated by the statistically significant interaction effect for soil organic matter content in the MANOVA. In general, upland LULC had higher soil organic matter content than the midland LULC. Apart from differences in LULC, this is mainly attributed to the influence of environmental factors on soil organic carbon concentration along altitudinal belts. Consistent with a study by Campos et al. (2013), the reduced soil organic matter decomposition rates as a result of lower temperatures in the upland part could contribute to the increase in the soil carbon. This finding lends strong support to implicate the interaction effects of LULC and (elevation) temperature on soil organic matter decomposition as discussed by Kirschbaum (1995), Garten Jr et al. (1999) and Griffiths et al. (2009). Even within the same AEZ unit, LULC significantly affects soil organic matter content. For example, a significant mean difference in soil organic matter content between soils under midland natural forest and midland cultivated land was observed. This indicates that the loss of soil organic matter could happen as a result of a remarkable decline in the addition of plant residues to the soil. This result is consistent with the results of various researchers (Lal, 2005; Lemenih and Itanna, 2004; Muñoz-Rojas et al., 2012; Yimer et al., 2007), who found a significant decline in soil organic matter content in the conversion of natural forest to cropland. Afforestation of barren land with *Eucalyptus* plantation can increase soil organic carbon content in the midland part. This is evident from the observed significant mean difference (mean = 2.66%) between soils under midland PF and midland BL. However, there was no evidence that this effect is true for the upland area as well. The upland cultivated land had significantly higher soil organic matter content than the midland cultivated land, suggesting differences in factors of soil organic carbon decomposition such as temperature and precipitation amount.

Soil compaction (bulk density)

Both LULC and altitude uniquely and significantly influenced bulk density. The highest significant difference in bulk density was observed in the conversion of natural woody vegetation to barren land. The conversion of natural forest to grassland in the Jedeb watershed was accompanied by significant changes in bulk density. Neill et al. (1997) also found a significant increase in bulk density after the conversion of forest to grassland. The compaction caused by cattle trampling in soils under grassland could be attributed to a significant higher bulk density. Therefore, one of the consequences of conversion of forest to grassland is soil compaction at least for the case of Jedeb watershed. Conversion to barren land also brings an increase in bulk density (compaction). According to Teferi et al. (2013) an area of 16 km^2 (5.4%) of the watershed was converted from forest to grassland and then conversion to barren land during the period 1957 – 2009. This means that about 5.4% had been degraded in terms of soil compaction. Thus, such types of unsustainable land use conversion need the attention of land users and land managers. Improved grazing land management has to be designed especially in the upland part to minimize overgrazing, which can lead to soil compaction.

6.5.2 Effects of Altitude on Soil Quality Indicators

Soil acidity (pH)

The non-significant effect of LULC and the significant effect of altitude on soil pH suggest that the variation in soil pH did not depend on land use types but on altitudinal variation. This result agrees with that of Bewket and Stroosnijder (2003) who found a non-significant difference in soil pH among different land use types in a nearby humid watershed. In contrast, Biro et al. (2013) found a significant difference in soil pH between land use types in semo-arid area of Sudan. The difference could be attributed to the difference in climatic condition between two areas. The mean soil pH of both the upland and the midland was less than 5, indicating the acidic nature of the soils of Jedeb watershed. At this pH value, availability of Phosphorus, Nitrogen, Potassium, Calcium, Magnesium, and Sulfur is reduced (McKenzie, 2004). This is largely attributed to the high rainfall (1400-1600 mm a^{-1}) resulting in leaching of exchangeable cations from the topsoil layer. The upland soil was significantly more acidic than the midland soil. Griffiths et al. (2009) also found significantly lower pH at higher elevations as compared to lower elevations. The problem of soil acidity may gradually lead to permanent soil degradation, unless the options of reducing the impact of soil acidity are designed. Maintaining the soil organic matter content can slow down the rate of acidification and prevent further leaching of important nutrients.

Bulk density

Upland soils had a significantly lower bulk density than the midland soils due to the apparent difference in the organic matter decomposition rates. Consequently, the upland area and the midland area of the watershed have different degrees of soil compaction. Thus, a distinction should be made based on AEZ during the selection of appropriate land management practices in the watershed. The midland part needs more efforts of reducing soil compaction than the upland part; i.e., what is best for the midland part may not work in the upland part.

Silt content (susceptibility to erosion)

The soils of the upland part of the watershed had significantly higher silt content than those in the midland part. The susceptibility of soils to erosion is usually related to the silt content given other factors of erosion are constant. Silt and fine sand are the least resistant to erosion (Richter and Negendank, 1977). Thus, the upland part of the watershed is more erodible than the midland part due to its significantly higher silt content. Soil management practices that enhance organic matter content of the soil could increase stability of the particles thereby increasing its resistance to detachment. However, excessive accumulation of organic matter might bring further lowering of soil pH (acidification). The other option to decrease upland soil's susceptibility to detachment is to keep the soil surface protected from raindrop impacts through the use of protective cover such as vegetation, mulch, and crop residues.

6.5.3 Effects of Land Use and Land Cover on Infiltration Rates

The relatively lower infiltration rate in soils under cultivated land could be attributed to the compaction as result of plough pan formation caused by the "*maresha*" plough (Biazin et al., 2011; Temesgen et al., 2009). Soils under grassland also show lower infiltration rate. This is largely explained by the effect of trampling by cattle on the grassland of the area (Taboada and Lavado, 1988). However, soils under natural woody vegetation depict higher infiltration rate. These marked differences are resulted from abundant litter cover, organic inputs, root growth and decay, and faunal activity in soils under natural vegetation, which significantly influence soil structure and soil water infiltration (Buytaert et al., 2005; Zhou et al., 2008).

6.6 CONCLUSIONS

The study showed that the application of multivariate techniques of PCA and MANOVA provides insights into the selection of SQIs. Organic matter content, bulk density, silt content, and soil pH were found to be appropriate SQIs to be used for evaluation of the sustainability of land use and management practices.

The impact of LULC on soil organic matter content was dependent on altitude, being higher in the upland LULC than the midland LULC. Additionally, the conversion of natural woody vegetation to cultivated land in the midland part decreases biomass productivity and reduces the quantity of biomass returned to the soil, and as a result decreases the soil carbon. The decline in soil organic carbon in the midland part might bring adverse effect on soil structure. Therefore, conversion of degraded or marginal lands to restorative land uses should be adopted to minimize further depletion of soil carbon in the midland part. For example, this study showed that afforestation of barren land with *Eucalyptus* plantation can increase soil organic carbon content in the midland part. However, there was no evidence that this effect is valid for the upland area as well.

The conversions of natural woody vegetation to grassland and to barren land were accompanied by significant changes in bulk density. The compaction caused by cattle trampling in soils under grassland could be attributed to a significant higher bulk density. Therefore, one of the consequences of conversion of forest to grassland is soil compaction at least for the case of the Jedeb watershed. Thus, such types of unsustainable land use conversion need the attention of land users and land managers. The upland area and the midland area of the watershed have different degrees of soil compaction. Thus, a distinction should be made based on AEZ during the selection of appropriate land management practices in the watershed. The midland part needs more efforts of reducing soil compaction than the upland part. Improved grazing land management has to be designed to minimize overgrazing which can lead to soil compaction.

Chapter 7

MONITORING OF SOIL MOISTURE IN RESPONSES

TO LAND USE AND LAND COVER CHANGES USING

REMOTE SENSING AND *IN-SITU* OBSERVATIONS[5]

7.1 ABSTRACT

An understanding of the impacts of land use and land cover change (LULCC) on soil moisture dynamics at different spatial scales is important for promoting effective management of water resources in an increasingly human-dominated biosphere. This study was carried out to understand spatiotemporal variability of soil moisture under different land cover change trajectories in the Upper Blue Nile (UBN) basin. The downscaled essential climate variable (ECV) surface soil moisture (SM) data set using a synergistic use of microwave-optical/infrared data was demonstrated to provide a reasonably good estimate of soil moisture at a resolution of 1 km. The root mean square residual (RMSR) ranges from 0.0164 to 0.0393 m^3 m^{-3} when using independent validation data sets of ECV SM. However, using independent validation data sets of Soil Moisture and Ocean Salinity (SMOS) SM, RMSR ranges from 0.0493 to 0.1807 m^3 m^{-3}. Both the downscaled and *in-situ* measured soil moisture were used to evaluate the impact of land cover change on soil moisture variation at the basin scale. Results of remote sensing observations indicated that there are significant differences in the mean soil moisture content between the reference trajectories (stable forest) and

---

[5] *This chapter is based on* Teferi, E., Wenninger, J., Uhlenbrook, S. (2015). Monitoring of soil moisture in responses to land use and land cover changes using Remote Sensing and *in-situ* observations (Intl J App Earth Observ Geoinform: under review).

other trajectories considered as determined by the ANOVA result (F(10,44) = 45.31, $p < 0.001$). Soils under the stable forest trajectories had the highest soil moisture levels. *In-situ* observations inside the Jedeb watershed also confirmed that the mean soil moisture levels under barren land and grassland are significantly different and lower from the mean soil moisture content of forest soils (F(4) = 33.44, $p < 0.001$). Thus, land cover changes, particularly forest cover changes have a significant influence on the soil moisture regime of the basin.

7.2 INTRODUCTION

Soil moisture is a key eco-hydrological variable controlling plant transpiration and photosynthesis with consequent impacts on the water, energy and biogeochemical cycles (Mahmood et al., 2011; Seneviratne et al., 2010). Soil moisture determines the partitioning of incoming solar radiation into sensible heat flux and latent heat flux and the partitioning of rainfall into surface runoff and subsurface infiltration and recharge. Thus, a better understanding of soil moisture variability and its controlling factors is crucial for hydrological research and applications, such as flood/drought prediction, climate forecasting and agricultural management practices etc. (Joshi and Mohanty, 2010; Starr, 2005).

Soil moisture is either measured directly using *in-situ* methods (e.g., gravimetric sampling, time domain reflectrometry (TDR)) (Grayson and Western, 1998; Mohanty et al., 1998), or estimated indirectly through remote sensing (RS) techniques. The remote sensing-based methods include active microwave (e.g.Ulaby et al., 1996), passive microwave (e.g.Njoku et al., 2003) and merged product (Liu et al., 2012a; Liu et al., 2011). The merged product is the Essential Climate Variable (ECV) soil moisture (ECV SM) data set. As *in-situ* measurements are labour intensive and limited to observations of small areas (Mohanty and Skaggs, 2001), the RS methods are preferred for estimation of soil moisture at larger spatial extents. However, the spatial resolution of remotely sensed soil moisture footprints (mostly 25 km) result in significant smoothing of the variability present in surface soil moisture (Western et al., 2004). This necessitates the use of downscaling algorithms to render the data suitable for various applications (Chauhan et al., 2003; Choi and Hur, 2012; Merlin et al., 2010).

Soil moisture exhibits a tremendous heterogeneity in space and time even in small catchments (Fernández et al., 2002; Gómez-Plaza et al., 2001). The variability of soil moisture results from the complex interaction of changes in land use and land cover (Fu et al., 2000;Hu et al., 2011;Sterling et al., 2013) and natural factors such as climate (Holsten et al., 2009) and topography (Burt and Butcher, 1985). Land use and land cover changes (LULCC) give rise to changes in soil physical properties and total evaporation to result in soil moisture change (Savva et al., 2013; Wang et al., 2012). Sterling et al. (2013) demonstrated the role of land cover change in altering the water cycle through direct changes to the timing and magnitude of total evaporation.

Afforestation and deforestation are the dominant type of LULCC in terms of both land area and hydrological impacts in most parts of the world (Calder, 1992). In Ethiopia, afforestation has received high priority in relation to sustainable land management (SLM) and climate issues. However, several studies have reported that forest impacts on the environment may not always be beneficial (Bruijnzeel, 2004; Calder, 1998; Sikka et al., 2003). Afforestation tends to increase soil moisture deficits in dry periods as compared to herbaceous plants because of the higher interception losses from forests in wet periods and increased transpiration losses in dry periods (due to deeper root systems). Bruijnzeel (2004) highlighted the increase in dry season flows following deforestation. In India, Sikka et al. (2003) identified reductions in low flows in the driest months of the year when comparing stream flows from a grassland catchment with that from a catchment afforested with *Eucaliptus globulus*. Replacing trees with herbaceous vegetation can increase surface water yields (Bosch and Hewlett, 1982; Farley et al., 2005).

However, afforestation will not necessarily increase soil moisture deficits and reduce dry season flows. Competing processes such as infiltration properties of soils and topography of a particular area may result in either increased or reduced soil water recharge (Bruijnzeel, 1990). Different research findings have been reported regarding the effects of LULCC on soil hydrology, largely due to the complexity of the competing processes affecting variation of soil moisture. Wahren et al. (2009) reported that afforestation causes an increased infiltration and soil water retention potential in Germany. Mapa (1995) demonstrated that reforested land had the highest steady infiltration rate and soil moisture retention in Kandy of Sri Lanka. In contrast, Yang et al. (2012) reported soil moisture deficits in all land covers changed with plantation in a semi-arid area of the Loess Plateau, China. Thus, effects of LULCC on soil water recharge or dry season flows are likely to be very site specific and this calls for investigations on the impact of LULCC on soil hydrology for a given environment before undertaking large scale afforestation or reforestation programs.

However, disentangling the impact of LULCC on soil moisture variation is difficult, as there are different competing controlling factors such as topography, rainfall, and soil type. Recently, Feng and Liu (2014) have tried to isolate the impact of deforestation and afforestation from other natural factors on soil moisture dynamics by comparing the soil moisture content of unchanged (stable) forest with other land cover trajectories. However, this study argues that in order to attribute the observed soil moisture variation to changes in land cover, the reference vegetation community and the other land cover trajectories need to be located in places where differences in topographic, geologic (soil) and climatic conditions are negligible. Thus, the objectives of the study are: (1) to downscale the merged microwave soil moisture product (ECV SM), and (2) to evaluate the impact of land cover change on soil moisture variation using the downscaled soil moisture and *in-situ* observations at the basin scale and watershed scale. The study was conducted in the Upper Blue Nile

(UBN) basin that forms the headwater for the Nile river. Thus, understanding the effects of land cover changes on soil moisture is of paramount importance.

7.3 MATERIALS AND METHODS

7.3.1 Description of Data Sets

ECV SM data set

The Essential Climate Variable (ECV) Soil Moisture (SM) data set (ECV SM version 0.1) is the output of blending the active and passive microwave soil moisture products, which are derived from SMMR (November 1978–August 1987), SSM/I (July 1987–2007), TMI (1998–2008), and AMSR-E (July 2002–2010) for the passive data sets, and ERS-1/2 (July 1991–May 2006), and ASCAT (2007–2010) for the active data sets. This data has been produced following the method as described by (Liu et al., 2012a;Liu et al., 2011). The homogenized and merged product presents surface soil moisture with a global coverage and a spatial resolution of 0.25°, based on the World Geodetic System 1984 (WGS 84) reference system, and the temporal resolution is 1 day. The data set spans over 30 years and covers the period from November 1978 to December 2010. The data set was obtained from European Space Agency (ESA) website (http://www.esa-soilmoisture-cci.org).

SMOS soil moisture product

The Soil Moisture and Ocean Salinity (SMOS) level 3 product was used for the purpose of validation of the predicted soil moisture. The SMOS level 3 (L3) product is processed, archived and distributed by the Centre Aval de Traitement des Données SMOS (CATDS). The CATDS SMOS L3 soil moisture product was obtained free of charge from CATDS website (http://catds.fr) for the year 2010. The Soil Moisture and Ocean Salinity (SMOS) satellite was launched in November 2009 by European Space Agency (ESA) in collaboration with the Centre National d'Etudes Spatiales (CNES) in France and the Centro para el Desarrollo Tecnologico Industrial (CDTI) in Spain. The SMOS satellite delivers surface soil moisture (0-5 cm) and brightness temperature products over terrestrial areas on a regular three-day basis with an accuracy of 0.04 m^3 m^{-3} with a spatiotemporal resolution of 30–50 km (Kerr et al., 2010). SMOS is different from the other satellites because it operates at a frequency of 1.4 GHz. Frequencies of 1–2 GHz (L-band) are ideal for soil moisture measurements, as L-band radiometers can penetrate the vegetation canopy and are more sensitive to soil moisture than higher frequencies over vegetated areas (Calvet et al., 2011).

MODIS products

Moderate Resolution Imaging Spectroradiometer (MODIS) data products such as vegetation index (MOD13A2), land surface temperature (MOD11A2), albedo (MCD43B3) were used for the purpose of assessing soil moisture. Besides, MODIS land cover product (MCD12Q1 Version 5) for the year 2001 and 2010 was used (Fig. 31). This data set contains the land covers classified from five different classification systems. This study selected the IGBP classification for its high classification accuracy (Friedl et al., 2010). The overall accuracy of this classification was 75%, with both the user's and the producer's accuracy being over 70% for most land cover classes (Friedl et al., 2010; Herold et al., 2008).

7.3.2 Downscaling of ECV SM

The general approach used in this study to estimate soil moisture at high resolution based on synergistic use of microwave and optical/infrared (IR) satellite data involves the following steps.

Step 1: Developing a linking model
A unique relationship was developed by Carlson et al. (1994) and Gillies and Carlson (1995) for the estimation of regional patterns of surface soil moisture availability with vegetation index (VI) versus land surface temperature (Ts) using a Soil Vegetation Atmosphere Transfer (SVAT) model. The theory can be referred to as the so-called "universal triangle" concept. A scatterplot of remotely sensed VI versus Ts often results in a triangular shape (Carlson et al., 1994; Price, 1990) or a trapezoid shape (Moran et al., 1994) if a full range of fractional vegetation cover and soil moisture contents is represented in the data. The emergence of the triangular (or trapezoid) shape in VI/Ts feature space is due to the relative insensitivity of Ts to soil water content variations over areas of dense vegetation, but increased sensitivity over areas of bare soil. The triangle is bounded by a "dry edge" and "wet edge". The dry edge is defined by the locus of points of highest temperatures that contain differing amounts of bare soil and vegetation and are assumed to represent conditions of limited surface soil water content and zero evaporative flux from the soil. Likewise, the wet edge represents those pixels at the limit of maximum surface soil water content.

Transforming the domain of the VI and Ts to the domain of scaled VI (N*) and scaled Ts (T*) serves to emphasize relative changes in the local VI signal over time while reducing the influence of atmospheric condition and satellite viewing angle (Gillies and Carlson, 1995). Thus, scaling the VI and Ts between the maximum and minimum values for pixels forming the triangular pattern allows one to compare images from different days and thereby allowing the monitoring of key parameters implicated in land-surface processes and land use and land cover change with time. The VI could be Normalized Difference Vegetation Index (NDVI), Enhanced Vegetation Index (EVI), or fractional vegetation cover (Fr). Since NDVI and EVI are dependent on the spatial resolution of the remote sensing data (Price, 1990), a more generalized VI, fractional vegetation cover (Fr), can be used. According to Gillies and Carlson (1995), Fr can be estimated as

$$Fr = (N^*)^2 \qquad [7.1]$$

where

$$N^* = \frac{EVI - EVI_{min}}{EVI_{max} - EVI_{min}} \qquad [7.2]$$

EVI_{min} and $EVI_{max,}$ correspond to the values of EVI for bare soil and a surface with 100% vegetation cover, respectively.

Similarly, a scaled land surface temperature (T*), which varies from 0 (T_{min}, the temperature pertaining to a dense clump of vegetation in well-watered soil) to 1 (T_{max}, the temperature of dry, bare soil – represented by the highest temperatures in the image); is be represented as

$$T^* = \frac{T - T_{min}}{T_{max} - T_{min}} \qquad [7.3]$$

The mathematical representation for surface soil moisture availability (M_0) from a scatter plot of scaled Ts and VI given by Eq. 7.4:

$$M_0 = \sum_{i=0}^{2} \sum_{j=0}^{2} \beta_{ij} T^{*i} V^j \qquad [7.4]$$

where V can be expressed either in terms of N* or Fr; T* is the scaled land surface temperature

Based on the universal triangle concept, Chauhan et al. (2003) disaggregated microwave soil moisture (M) into higher resolution soil moisture using Ts, VI, and surface albedo (A). Thus, Eq. 7.4 is modified to

$$M = \sum_{i=0}^{n} \sum_{j=0}^{n} \sum_{k=0}^{n} \beta_{ijk} T^{*i} V^j A^{*k} \qquad [7.5]$$

where A* = (A-A_{min})/ (A_{max}-A_{min})

In order to avoid multicollinearity problem in the multiple regression, Eq. 7.5 can be expanded as

$$M = \beta_0 + \beta_1 T^* + \beta_2 V + \beta_3 A^* \qquad [7.6]$$

Step 2: Integrating spatial variability into the model.

Spatial data exhibits two properties that make it difficult to meet the assumptions and requirements of non-spatial statistical methods, such as ordinary least square (OLS) regression: (1) the spatial autocorrelation; and (2) the non-stationarity characteristic of spatial data. In this study, spatial regression methods were applied to robustly manage these two characteristics of spatial data and to improve their ability to model data relationships using spatial statistics tools in ArcGIS software. Spatial autocorrelation between observations is non-trivial when constructing the relationship between microwave soil moisture and MODIS derived variables (Ts, EVI, A). Therefore, it is important to adjust for spatial autocorrelation associated with geographic data (Ji and Peters, 2004). However, many studies using MODIS data for downscaling of soil moisture have ignored these effects. Ignorance of spatial autocorrelation can result in incorrect conclusions to be reached regarding the research hypotheses.

The Koenker's studentized Bruesch-Pagan statistic (KBP) was applied to determine if the modelled relationships either change across the study area (non-stationarity) or vary in relation to the magnitude of the variable to be predicted (heteroscedasticity) (Koenker, 1981). The Jarque-Bera statistic indicates whether or not the residuals are normally distributed. The null hypothesis for this test is that the residuals are normally distributed (Jarque and Bera, 1980). When this test is significant, the residuals are not normally distributed, indicating that the model is biased. The Global Moran's I statistic was calculated on the regression residuals to test for statistically significant spatial autocorrelation (Moran, 1950).

Step 3: High resolution soil moisture estimation

To estimate soil moisture at a 1 km resolution, first, a spatial regression model was set up between ECV SM and aggregated EVI, aggregated albedo, aggregated Ts using Eq. 7.6. Second, the regression coefficients β_{ijk} were determined. Then, these regression coefficients were applied on, high-resolution EVI, Ts, and albedo to obtain soil moisture values at a high resolution (Fig. 29). In order to measure how well the regression equations can predict on independent samples, validation was performed based on both ECV SM and SMOS SM observations. The sample was randomly divided into a calibration and validation set. The regression models were developed on calibration samples and these equations were then applied to the validation samples. The Root Mean Square Residual (RMSR) between the regression-derived soil moisture and the SMOS-derived soil moisture was then calculated.

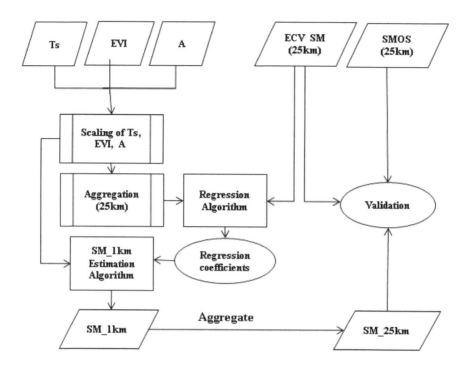

Figure 29 Flowchart for downscaling of the ECV SM product (25 km) to a 1 km resolution. The variables are defined in Section 7.3.2. Ts (land surface temperature), EVI (Enhanced Vegetation Index) and A (Albedo).

7.3.3 *In-situ* soil moisture measurements

Soil moisture content was measured in the field using a *Theta probe* soil moisture sensor (ML2x), which responds to changes in the apparent dielectric constant of the medium (Gaskin and Miller, 1996). However, a calibration was necessary because the soil's apparent dielectric constant depends not only on the water content, but also on factors such as bulk density, porosity and chemical composition of the soil water for a specific soil. Thus, a soil specific calibration was carried out with volumetric soil water contents determined from 21 soil samples. The regression equation is $Y=0.9092X+0.0337$, $R^2=0.9271$, where X is the measured soil moisture content using the *ThetaProbe* ML2x, and Y is the volumetric soil water content (m^3 m^{-3}) determined from gravimetric/bulk density measurements. All the subsequent field measurements were converted using this equation. Soil moisture measurements were taken on May 01, 2010 at two depths: 0-5cm, and 15-20cm. A soil auger was used to dig the soil up to the anticipated soil depth (Fig. 30 a & b). Considering the degree of variability and the destructive nature of the soil auger method, eight random probe locations within a 2 m circle per sample plot of land cover were selected for measurement. The following land cover types with a land use history of at least 23 years were selected:

Forest, *Eucalyptus* plantation, grassland, cropland, and barren land. The crop type on the cultivated land was *engido* (*Avena spp.*). The selected type of forest plantation was *Eucalyptus globulus*, as it is a widely planted tree in the study area. In order to minimize the soil moisture variation caused by other factors, sampling sites were carefully chosen in such a way that all sites had similar soil types (i.e. Leptosls), slope class (i.e. 15-30%) and altitude range (2400-2500 m a.s.l.).

Figure 30 Pictures of in-situ soil moisture observation using Theta Probe ML2x over Sentera grassland (a & b)

7.3.4   Impact of land cover change on soil moisture changes

MODIS based land cover trajectories of two time periods (2001 and 2010) were developed by overlaying both land cover maps (Fig. 31). A trajectory code with two digits can be obtained. The first and the second digits represent the land cover type in the year 2001 and 2010, respectively. The dominant trajectories were retained for soil moisture change analysis.

If soil moisture change in the permanent forest trajectory is assumed to be a representation of natural influences, it can be used as a control group to evaluate soil moisture changes in response to land cover change trajectories. Before taking samples, the basin was first subdivided into spatial elements of homogeneous areas with respect to factors influencing on soil moisture based on major soil types and topography (altitude and slope). Altitude also provides a useful proxy for the climate condition of places in Ethiopia (Hurni, 1999). By intersecting the three layers, one dominant sampling unit was selected. This unit is characterized by an altitude range between 1300 and 2600 m a.s.l (i.e. "Weyna Dega" or moderate agro-climatic zone), a soil type of Haplic Alisols, and a slope range of less than 3%. One-way analysis of variance (Holsten et al.) was then carried out to evaluate the relationship between soil moisture level under stable forest trajectories and other land cover change trajectories. An Anderson-Darling (AD) test (Stephens, 1974) was performed to test the normality assumption. Since the purpose of this research is to compare the mean of each of several land cover trajectories with the mean of reference trajectory (stable

112

forest), Dunnett's C test was performed to see which group means are significantly different from the control group mean. It can be used for equal and unequal group sizes where the variances are unequal (Toothaker, 1991).

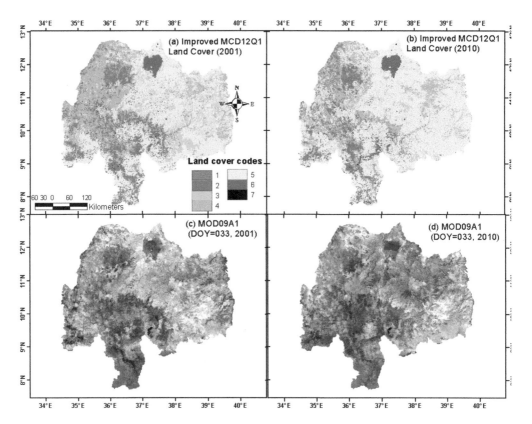

Figure 31 Improved MCD12Q1 land cover classes for 2001 (a), 2010 (b), Band2, Band1, Band3 combination of MODIS surface reflectance image for 2001 (c), and (d) (2010)

7.4 RESULTS AND DISCUSSION

7.4.1 Downscaling ECV Surface Soil Moisture

Spatial regression diagnostics result

Table 28 shows the regression diagnostics result. Results from the spatial autocorrelation test on the regression residuals using the Global Moran's I statistic indicates that the sampling points used in the model development are randomly distributed, as the p-value is not statistically significant. Thus, it is possible to trust the calibrated model. A positive Moran's I index value indicates tendency towards clustering while a negative Moran's I index value indicates tendency towards dispersion. The Jarque-Bera tests for all models are not statistically significant, indicating that the residuals from a regression model are normally distributed and the models are not biased. Furthermore, the relationships between the explanatory variables and the dependent variable are stationary, as indicated by the non-statistically significant Koenker tests. This means, that the predicting capability of the predictors is location invariant.

Table 28 Spatial regression diagnostics for non-normality, non-stationarity, and spatial autocorrelation

DOY	Koenker (BP) test	Jarque-Bera test	Global Moran's I test
25	2.48 (p = 0.48)	3.85 (p = 0.15)	0.10 (p = 0.22)
33	3.68 (p = 0.30)	3.85 (p = 0.15)	0.02 (p = 0.72)
41	2.43 (p = 0.49)	2.99 (p = 0.22)	0.02 (p = 0.59)
49	4.36 (p = 0.23)	3.76 (p = 0.15)	0.08 (p = 0.27)
57	3.05 (p = 0.38)	5.30 (p = 0.07)	0.08 (p = 0.24)
65	2.32 (p = 0.51)	4.46 (p = 0.11)	-0.09 (p = 0.47)
73	5.77 (p = 0.12)	3.41 (p = 0.18)	0.024 (p = 0.61)
81	2.54 (p = 0.47)	2.71 (p = 0.26)	0.14 (p = 0.13)
89	7.09 (p = 0.07)	3.26 (p = 0.20)	0.03 (p = 0.61)
97	2.67 (p = 0.45)	3.30 (p = 0.19)	0.12 (p = 0.19)
105	2.97 (p = 0.40)	2.56 (p = 0.28)	0.05 (p = 0.52)
113	2.06 (p = 0.56)	1.58 (p = 0.45)	-0.15 (p = 0.12)
121	2.92 (p = 0.40)	3.93 (p = 0.14)	-0.04 (p = 0.83)
129	5.18 (p = 0.16)	2.24 (p = 0.33)	-0.11 (p = 0.30)

* When Koenker (BP) Statistic test is not statistically significant (p > 0.05), the relationships modeled are consistent (either due to stationarity or homoskedasticity); when Jarque-Bera Statistic test is not statistically significant (p > 0.05) model predictions are biased (the residuals are normally distributed); when the Global Moran's I Statitic is not statistically significant (p < 0.05), the spatial distribution of feature values is the result of random spatial processes.

7.4.2 Model Calibration and Validation

Table 29 shows the calibration and validation result. Validation of soil moisture estimation results is difficult mainly because of the practical problems in undertaking *in-situ* soil moisture measurements that represents the soil moisture in the footprints of satellite soil moisture sensors. Moreover, since surface soil moisture changes very rapidly over time and space, the average soil moisture computed from point measurements within a footprint does not give an accurate representation of the soil moisture in the footprint. In view of these uncertainties, validation was performed based on both ECV SM and SMOS SM observations using data splitting technique. Regression models provided a good fit with an R^2 and RMSR ranging from of 0.61 to 0.95 and 0.0164 to 0.0393 m^3 m^{-3}, respectively, using independent validation data sets of ECV SM. The results are in agreement with that of Chauhan et al. (2003)'s RMSR of 0.016 m^3 m^{-3} and Ray et al. (2010)'s RMSR of 0.017 m^3 m^{-3}. However, using independent validation data sets of SMOS SM, regression models provided a moderate fit with an R^2 and RMSR ranging from of 0.13 to 0.61 and 0.0493 to 0.1807 m^3 m^{-3}, respectively. Chauhan et al. (2003) obtained RMSR ranging from 0.131 to 0.179 m^3 m^{-3}, and correlation coefficients ranging from 0.127 to 0.725 in validation of AMSR-E soil moisture using *in-situ* measurements.

Table 29 Calibration and validation result

DOY	β_0	β_1	β_2	β_3	n	Calibration (50%)		Validation (50%)		Validation (SMOS)	
						R^2	RMSR	R^2	RMSR	R^2	RMSR
025	**1.24**	**-1.16**	**0.32**	**-0.1**	60	0.78	0.0246	0.72	0.0261	0.36	0.0927
033	**2.31**	**-2.3**	**0.06**	0.02	63	0.85	0.0300	0.77	0.0338	0.21	0.0676
041	**1.67**	**-1.61**	-0.02	0.03	57	0.82	0.0400	0.64	0.0393	0.17	0.0622
049	**0.87**	**-0.69**	**0.27**	**-0.19**	62	0.71	0.0249	0.61	0.0245	0.29	0.0635
057	**2.20**	**-2.13**	0.08	**-0.07**	61	0.82	0.0212	0.79	0.0235	0.29	0.0506
065	**2.65**	**-2.56**	-0.04	**-0.06**	57	0.93	0.0175	0.93	0.0181	0.28	0.0713
073	**2.58**	**-2.5**	0.23	**-0.16**	52	0.86	0.0266	0.83	0.0325	0.52	0.0740
081	**2.55**	**-2.39**	-0.1	**-0.10**	51	0.92	0.0188	0.88	0.0212	0.30	0.0826
089	**2.48**	**-2.29**	-0.03	**-0.20**	58	0.95	0.0151	0.94	0.0195	0.43	0.0998
097	**2.95**	**-2.81**	**-0.17**	-0.05	56	0.93	0.0236	0.91	0.0215	0.42	0.0872
105	**2.59**	**-2.43**	-0.02	**-0.12**	57	0.97	0.0166	0.95	0.0189	0.61	0.0746
113	**1.88**	**-1.69**	0.09	**-0.11**	59	0.87	0.0200	0.80	0.0264	0.13	0.0493
121	**2.13**	**-1.89**	-0.07	**-0.08**	53	0.91	0.0158	0.89	0.0164	0.18	0.1807
129	**2.28**	**-2.11**	**0.13**	**-0.10**	57	0.96	0.0153	0.95	0.0166	0.55	0.0885

*coefficients written in bold are significant at the 0.05 level

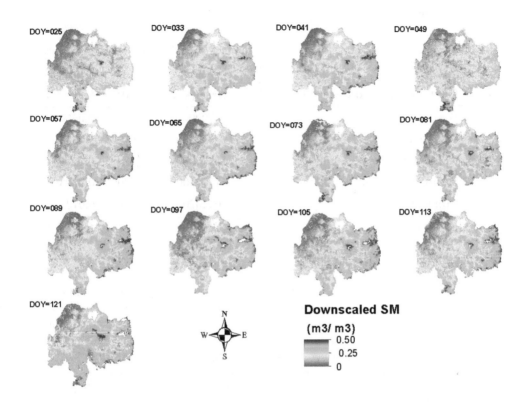

Figure 32 The spatio-temporal patterns of the downscaled soil moisture for the dry season of 2010 (from Day of Year (DOY) 025 to DOY 121)

7.4.3 Effects of Land Cover Change on Downscaled Soil Moisture Dynamics

The spatio-temporal pattern of the downscaled soil moisture for the dry season of 2010 is shown in Fig. 32. Land surface temperature and albedo were found to be significant predictors of soil moisture throughout the dry season, whereas enhanced vegetation index was able to significantly predict the surface soil moisture only on a few 8-day composite periods. Surface soil moisture is negatively related to both LST and A.

Fig. 31 shows that the basin is a cropland dominated landscape in both years. Table 30 summarizes the surface soil moisture contents (mean ± standard deviation) under different land use and land cover types for the year 2010. Surface soil moisture content was much higher in forested areas than in any other land cover type, indicating a lower level of evaporation from the soil and generally good soil moisture status under the forests soils.

Table 30 Average values and standard deviations for surface soil moisture contents ($m^3 m^{-3}$) under different land use and land cover types

Code	Land use and land cover types	Area (km^2)	SMC (Mean±STD)
1	Forests	4579000	0.3000±0.0556
2	Woodland	23057000	0.1966±0.0538
3	Grassland	19496000	0.2321±0.087
4	Shrubs and bushes	47770000	0.2065±0.0703
5	Cropland	106437000	0.2731±0.0731
6	Water bodies	3217000	-
7	Urban and Barren	189000	-

SMC=Soil Moisture Content, STD = Standard Deviation

Considering that soil moisture variation under soils of the reference land cover trajectory reflects the natural condition (i.e. virgin soil), the differences between the reference and other trajectories would be the contribution from land cover change. Stable forest was taken as a primary reference land cover trajectories. This trajectory represents the natural forest that remains unchanged between 2001 and 2010. An Anderson-Darling (AD) test of normality confirmed that the data is normally distributed (AD=0.70, $p > 0.05$) and, thus, the data meets the requirement for running ANOVA. The test of homogeneity of variance was found to be significant; indicating that the assumption of equality of variance has not been met. Thus, the Dunnett's C test was applied for multiple comparison of means.

Table 31 Dunnett's C multiple comparison test for mean difference in soil moisture between reference (stable forest) and other trajectories

Reference trajectories	Land cover change trajectories	Mean Difference in soil moisture ($m^3 m^{-3}$)	Std. Error
Stable forest (0.3638 $m^3 m^{-3}$)	Forest to cropland (0.3062 $m^3 m^{-3}$)	0.0576	0.0111
	Stable woodland (0.2477 $m^3 m^{-3}$)	0.1162*	0.0102
	woodland to shrubs and bushes (0.1920 $m^3 m^{-3}$)	0.1718*	0.0112
	Woodland to cropland (0.2093 $m^3 m^{-3}$)	0.1545*	0.0160
	Stable grassland (0.2302 $m^3 m^{-3}$)	0.1336*	0.0126
	Grassland to cropland (0.2158 $m^3 m^{-3}$)	0.1480*	0.0111
	Shrubs and bushes to grassland (0.1734 $m^3 m^{-3}$)	0.1904*	0.0112
	Stable shrubs and bushes (0.1876 $m^3 m^{-3}$)	0.1763*	0.0107
	Shrubs and bushes to cropland (0.1630 $m^3 m^{-3}$)	0.2008*	0.0101
	Stable cropland (0.2327 $m^3 m^{-3}$)	0.1312*	0.0221

*The mean difference is significant at the .05 level.

There were statistically significant differences in soil moisture contents between stable forest and other trajectories as determined by one-way ANOVA ($F(10,44) = 45.31$, p

117

< 0.001). This may imply that land cover change have a significant influence on the soil moisture regime of the basin. In contrast, Venkatesh et al. (2011) obtained a non-significant impact of land cover types on soil moisture in India, reflecting that other factors might have strong influence on the soil moisture variation. Table 31 depicts the Dunnett's C multiple comparison test result, indicating which specific group means are significantly different from the control group mean (stable forest). The soils under stable forest trajectories had the highest soil moisture level (i.e.0.36 m³ m⁻³). One of the explanations for this could be that soils under natural forests tend to be relatively porous because trees loosen the soil and accumulate more organic matter, thereby increasing the water retention capacity in forest sites. Zhang et al. (2011b) found greater soil water retention capacities for forest soils than for soils under shrubs. Thus, removal of these forests could cause a loss in soil moisture storage capacity. For example, statistically significant ($P < 0.001$) highest declines in soil moisture levels were observed due to conversion of shrubs and bushes to cropland (0.20 m³ m⁻³) and to grassland (0.19 m³ m⁻³) (Table 31). Such conversions can affect soil properties such as bulk density, porosity, and organic matter content, and hence changes the water storage capacity of the soils underneath.

The soil moisture under a forest to cropland conversion did not have a significant difference to that of the stable forest (Table 32). The most likely explanation for this could be because of the recent conversion from forest, the soil did not get sufficient time to significantly change its properties from its original state. Thus, the age of conversion might contribute to the insignificant difference in soil moisture levels with reference to stable forest. Figure 33 clearly shows that the soil moisture under a forest to cropland converted area had lower soil moisture levels than that of stable forest throughout the dry season.

The soil moisture level under a stable woodland trajectory was significantly different from that of stable forest trajectory ($p < 0.05$). However, its mean difference (i.e.0.12 m³ m⁻³) with soil moisture content of stable forest is small as compared to the mean difference of other trajectories. This means, if woodlands are protected from being deforested, it is possible to get nearly equivalent benefits of natural forest with regard to soil moisture management. Converting woodland to bushes and shrubs can cause an equivalent decline in soil moisture levels (0.17 m³ m⁻³) than it would be in woodland to cropland conversion (0.16 m³ m⁻³).

The temporal soil moisture variations with respect to the land cover trajectories are shown in Figure 33. The soil moisture levels of soils under a stable forest trajectory and shrubs and bushes to cropland trajectory remained highest and lowest, respectively, throughout the dry season in the year 2010. The soil moisture under soil of all land cover trajectories reach their minimum at DOY=73 (March 07-14). The soil moisture difference among land cover trajectories is well distinct between the period DOY 89 and DOY 121 (i.e. March 23 – May 01).

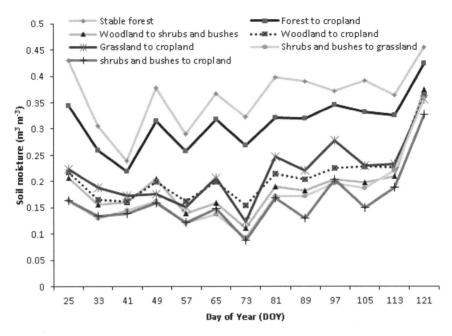

Figure 33 The temporal variations of mean soil moisture contents of sampled areas

7.4.4 Assessing the Effects of Land Cover Change on Soil Moisture Using *In-situ* Observations

The interaction effect of land cover and depth was not statistically significant ($p >$ 0.05) (Table 32). In contrast, Yimer et al. (2008) reported a significant integration effect of land use and soil depth on soil moisture variation elsewhere in Ethiopia. There were statistically significant differences in soil moisture content between forest land cover and other land cover types as determined by two-way ANOVA ($F(4) = 33.44$, $p < 0.001$). As with the downscaled soil moisture, the in-situ measurements also confirmed that the mean soil moisture level under forest was significantly higher than that of other land cover types. Yimer et al. (2008) reported higher soil moisture contents of forest soils in Bala Mountains in Ethiopia. The mean soil moisture content at different depths was also significantly different ($F(1) = 9.74$, $p < 0.01$). Generally, the mean soil moisture content increases with increasing soil depth. However, this difference in soil moisture is negligible under forest soils. This could be because of the reduced soil evaporation resulting from a dense ground cover. Highest spatial variability in mean soil moisture content is observed under grassland soils, as reflected by a standard deviation of 0.048 m^3 m^{-3} for the top soil layer and 0.041 m^3 m^{-3} for the bottom layer.

Table 32 ANOVA test result for the impacts of land cover and soil depth on soil moisture

Land use and land cover types	Sampling Depth (cm)	Soil moisture content (m^3 m^{-3})	
		Mean	Std. Deviation
Forest	0-5	0.2173	0.0402
	15-20	0.2158	0.0179
Eucalyptus plantation	0-5	0.1106	0.0203
	15-20	0.1350	0.0296
Grassland	0-5	0.0943	0.0480
	15-20	0.1282	0.0411
Cropland	0-5	0.1103	0.0288
	15-20	0.1410	0.0365
Barren land	0-5	0.0745	0.0406
	15-20	0.1045	0.0177

Two-way ANOVA test
Land Cover (LC), $df=4$, $F =33.48$, $p < 0.001$
Depth (D), $df=1$, $F =9.74$, $p =0.003$
LC X D, $df=4$, $F =0.73$, $p =0.575$

Statistically significant ($P < 0.001$) highest deviation from the mean soil moisture content of forest soils were observed in barren land (0.13 m^3 m^{-3}) and grassland (0.12 m^3 m^{-3}) (Table 33). This indicates that deforestation could reduce the ability to hold and absorb soil moisture because of the loss of organic matter and an increase in bulk density (Lepsch et al., 1994). According to Teferi et al. (2013), 15.5% of the Jedeb watershed was covered by natural woody vegetation in 1957, but in 2009 only 3.9% remained. Thus, the prevailing deforestation in the watershed might bring increased surface runoff resulting from a reduced soil moisture holding capacity. Forest to grassland conversation caused a reduction of 0.12 m^3 m^{-3} mean soil moisture. This could be related to the compaction of soils caused by cattle trampling. The size of grazing land left after conversion to cropland is very small in the case study site (Teferi et al., 2013). Cattles are now grazing on those few grazing areas and this creates pressure on the existing grasslands. Murty et al. (2002) also reported an increase in bulk density as a result of forest to grassland conversion.

Although a significant difference in mean soil moisture level is observed between forest and *Eucalyptus* plantation, it is the smallest mean difference (i.e.0.04 m^3 m^{-3}). This means that next to natural forest, *Eucalyptus* plantation can be used to restore the barren land to soils with relatively good soil moisture holding capacities. As a type of revegetative measure, *Eucalyptus* plantation may be the best choice for maintaining good soil moisture states of the soil. However, log-term observation is required to explore the practical impact of afforestation on soil moisture dynamics.

Table 33 Mean differences in soil moisture contents in soils of different land cover types in relation to forest soils

Reference land cover	Other land cover types	Mean Difference in soil moisture ($m^3 m^{-3}$)	Std. Error
Forest	*Eucalyptus* plantation	0.0370*	0.0191
	Grassland	0.1152*	0.0191
	Cropland	0.0908*	0.0191
	Barren land	0.1270*	0.0191

*. The mean difference is significant at the .001 level.

7.5 CONCLUSIONS

This study was carried out to understand spatiotemporal variability of soil moisture under different land cover change trajectories in the UBN basin located in Ethiopia. Since the spatial resolution of the already available space-borne soil moisture observations is much greater than the typical scale of soil moisture variability, high-resolution soil moisture was estimated through a synergistic use of microwave-optical/infrared data. The 'universal triangle' concept was used to establish relations between ECV SM and optical/infrared (land surface temperature, EVI, and albedo) using a spatial regression technique that adjusts for spatial autocorrelation inherent in spatial data. Through comparisons to SMOS surface soil moisture, the downscaled multi-satellite surface soil moisture dataset (ECV SM) was demonstrated to provide a reasonably good estimate of soil moisture at 1 km resolution in the basin. Although such downscaling procedure provides some improvement, better methods that can downscale ECV SM data set to finer than 1 km spatial resolution, are still needed.

Both the downscaled and *in-situ* measured soil moisture values were used to evaluate the impact of land cover change on soil moisture variation at the basin scale. Results indicated that there are significant differences in the mean soil moisture contents between the reference trajectories (stable forest) and other trajectories considered as determined by the ANOVA results. The soils under stable forest trajectories had the highest soil moisture level. *In-situ* observations from the case study Jedeb watershed also confirmed that the mean soil moisture levels under barren land and grassland significantly deviate highly from the mean soil moisture content of forest soils. Thus, removal of these forests could cause a loss in soil moisture storage capacity of soils. Significantly highest declines in soil moisture levels were observed due to conversion of shrubs and bushes to cropland and to grassland. The observed mean difference in soil moisture between stable woodland and stable forest was small as compared to the mean difference of the other trajectories. This means, if woodlands are protected from being deforested, it is possible to get nearly equal benefits of natural forest with regard to soil moisture management. The study concludes that land cover changes have a significant influence on the soil moisture regime of the basin.

Chapter 8

PREDICTING SOIL WATER RETENTION

CHARACTERISTICS IN HIGH ALTITUDE TROPICAL

SOILS[6]

8.1 ABSTRACT

Predicting soil water retention characteristics from readily available soil data using pedotransfer functions (PTFs) is crucial for simulation of water and solute fluxes in the unsaturated zone. Due to distinctive pedological properties of high altitude tropical soils, PTFs developed elsewhere cannot be applied directly to soils of the Upper Blue Nile/Abay region. The objectives of this study were to: (i) derive and verify PTFs for the parameters of the van Genuchten model and for water contents at specific water potentials using the *all possible subset* regression method; (ii) analyse the predictive ability of Artificial Neural Network (ANN) and k-nearest neighbour (k-NN)-based point PTFs for the estimation of water retained at -330 cm(θ_{FC}) and -15000 cm(θ_{PWP}); and (iii) derive class PTFs and map the Available Water Content (AWC) at the basin scale. Results showed a mean root mean squared difference (MRMSD) of 0.0349 cm^3cm^{-3} for the proposed point PTFs and a MRMSD of 0.0508 cm^3cm^{-3} for the proposed continuous PTFs, compared with MRMSD of 0.0996 cm^3cm^{-3} for PTFs of Hodnett and Tomasella (2002). The addition of bulk density to the particle size distribution (PSD) markedly improved the accuracy of both k-NN and

---

[6]*This chapter is based on* Teferi, E., Wenninger, J., Uhlenbrook, S. (2015). Predicting soil water retention characteristics in high altitude tropical soils – A case study from the Upper Blue Nile/Abay, Ethiopia (CATENA: under review).

ANN PTFs in predicting θ_{PWP}, but not in estimating θ_{FC}. In estimating both θ_{FC} and θ_{PWP}, the addition of organic matter content to PSD increased the accuracy of k-NN PTFs, but not that of ANN PTFs. The difference in accuracy between ANN and k-NN methods is dependent on the type of the predicted parameters (θ_{FC} or θ_{PWP}). The PTFs of k-NN performed better than those of ANN in estimating θ_{FC}, whereas the two methods were not significantly different in predicting θ_{PWP}. A map of the total AWC through the use of the derived class PTFs was established for the whole Upper Blue Nile/Abay basin (199, 812 km^2). Thus, collecting and integrating fragmented available soil survey data enables scientists and planners to develop area-specific PTFs and generate new information such as maps of available water at the basin scale.

8.2 INTRODUCTION

Modern analyses of hydrological processes such as simulation of water and solute fluxes rely heavily on proper understanding of soil water retention characteristics (SWRC). Accurate characterization of SWRC requires a large number of samples due to high temporal and spatial variability in the hydraulic characteristics. This again demands expensive and time consuming field and laboratory work. It is therefore imperative to predict data we need (soil hydraulic properties) from what we have (readily available soil data) using pedotransfer functions (PTFs) (Bouma, 1989).

In general, PTFs can be classified into two main groups: point and parametric PTFs. Point PTFs predict the soil water content at a specific soil water pressure head (Obalum and Obi, 2013; Rawls et al., 1982; Tomasella et al., 2003). Parametric PTFs estimate the empirical parameters of functions that describe the dependence of the water content on the soil water potential (e.g.,Rawls and Brakensiek, 1989; Schaap et al., 1999; Vereecken et al., 1989). Parametric PTFs assume that the hydraulic properties can be described adequately with hydraulic models, such as van Genuchten (1980). The parametric PTFs again can be categorized into class PTFs and continuous PTFs. Class PTFs predicts the average hydraulic characteristics for a soil texture or pedological class (e.g.,Pachepsky and Rawls, 2003; Wösten et al., 1995). Continuous PTFs are a series of equations used to predict water retention model parameters from for instance particle size distribution (PSD), organic matter (OM), bulk density (BD), pH and Cation Exchange Capacity (Wösten et al., 2001).

Methods used to derive PTFs can be classified into statistical regression methods (Vereecken and Herbst, 2004) and data mining and exploration methods (Pachepsky and Schaap, 2004). For statistical regression methods, model (variable) selection is not straightforward. However, for the data mining and exploration techniques such as Artificial Neural Network (ANN) and k-Nearest Neighbour (k-NN), selection of predictors or equations can be automated and there are no predefined mathematical functions. Several studies have used ANNs for PTF development (e.g.Børgesen et al.,

2008; Schaap and Leij, 1998a). In general, ANN PTFs performed well in terms of root mean square error (Vereecken et al., 2010). The performance of k-NN technique is comparable with that of ANN. Despite their good performance, data mining techniques are very data demanding and their application requires building a large data base.

It has been well recognized by the scientific community that building larger soil data bases by combining fragmented available data can facilitate analysis of soil hydraulic properties. As a result several large databases have been developed over the past decades, such as, Unsaturated Soil Hydraulic Database (UNSODA) (Leij et al., 1996) and Hydraulic Properties of European Soils (HYPRES) (Wösten et al., 1999). Furthermore, a myriad of PTFs have been derived and tested for the USA (e.g.,Rawls et al., 1982) and Europe (e.g.,Bruand et al., 2003; Vereecken et al., 1989;Wösten et al., 1999) based on these large databases. However, soils of tropical regions are largely underrepresented and only a limited number of samples are available in such databases. This resulted in increased inability and uncertainty of PTFs derived from temperate soils in estimating the hydraulic properties of tropical soils (Young et al., 1999). Due to distinctive pedological properties of tropical soils, Hodnett and Tomasella (2002) emphasized the need to develop separate PTFs. Many studies have demonstrated that the performance of a PTF depends on the data sets on which it is developed and tested (e.g.,Cornelis et al., 2001; Schaap and Leij, 1998b; Young et al., 1999). Therefore, it is important to develop and test a new PTF for tropical soils especially for those located in high altitude temperate climates, as these soils were not addressed by previous tropical PTFs such as the PTFs of Hodnett and Tomasella (2002). Recognizing that data availability for tropical soils is limited, it was necessary to build a new database for development and validation of new PTFs by compiling soil hydraulic data from different sources. This paper focuses on developing and testing new pedotransfer functions that are specific to soils of the Upper Blue Nile/Abbay region using newly collected or compiled data. The results of this study are very important from context of Upper Blue Nile basin for two reasons. First, the soils of Upper Blue Nile have not been studied for their SWRC and hence, information on parametric function that describes SWRC of these soils is not known. Second, Upper Blue Nile basin is the major water source area of the Nile river.

The objectives of this study were to: (i) derive and verify point, continuous, and class PTFs of soil water retention for soils of the Upper Blue Nile/Abay region; (ii) evaluate the effectiveness of Artificial Neural Network (ANN) and k-nearest neighbour (k-NN) point PTFs for the estimation of water contents at field capacity and permanent wilting point; and (iii) estimate and map the Available Water Content (AWC) for the Upper Blue Nile/Abay river basin.

8.3 MATERIALS AND METHODS

8.3.1 Description of Data Sets

Comprehensive data on soil properties varying in their complexity of information were collected and compiled from different projects of the Ministry of Water and Energy (Table 34). The study consists of three types of data sets: *data set 1*; *data set 2*; and *data set 3* (Fig. 34). Figure 35 shows the textural distribution of all the data sets on the USDA textural triangle. The soil sample analyses were carried out by the Ministry of Water Resources Laboratory and National Soils Laboratory of Ethiopia. Outliers were analyzed for quality control and consistency by drawing Box-and-Whisker plots and samples that are beyond the quartiles by one and a half times the inter quartile ranges were considered as potential outliers and excluded. The location of the data sets represented most of the major soil groups in the Upper Blue Nile/Abay basin. Each data set was split into two groups: (i) calibration data (70%), and (ii) testing data (30%).

Table 34 Description of data sets used in this study: Particle size distribution data (PSD) (i.e. sand (%), silt (%), and clay (%)), bulk density (BD) (g cm^{-3}), organic carbon (OC) (%), pH (-), Cation Exchange Capacity (Dinku et al.) (cmol kg^{-1}), and volumetric water content (VWC) at different matric potentials.

Location (Sub-basins)	n	Water content at selected matric potentials							Physical and chemical data				
		pF0	pF2.3	pF2.5	pF3	pF3.5	pF3.7	pF4.2	BD	PSD	OC	pH	CEC
Data set 1													
Upper Beles	53			*	*	*	*	*	*	*	*	*	*
Lake Tana	26			*	*	*	*	*	*	*	*	*	*
South Gojjam	44	*	*	*	*		*	*	*	*	*	*	*
Data set 2													
Anger	87		*					*	*	*	*	*	*
Beshilo	6		*					*	*	*	*	*	*
Didessa	24		*					*	*	*	*	*	*
Guder	8		*					*	*	*	*	*	*
Lake Tana	414		*					*	*	*	*	*	*
South Gojjam	44		*					*	*	*	*	*	*
Upper Beles	114		*					*	*	*	*	*	*
Data set 3													
Abay basin	890									*			

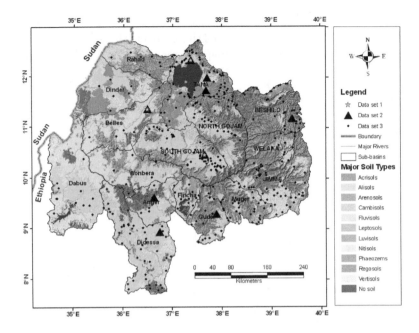

Figure 34 Map of study are showing the locations of samples and distribution of soil types in the Upper Blue Nile/Abay basin (Source: BCEOM, 1998b)

Data set 1 consists of 45 samples of newly collected data (South Gojjam sub-basin) and compiled data from previous projects (Upper Beles and Lake Tana sub-basins). This data set was used to develop and test point, continuous and class PTFs. A total of 44 undisturbed soil samples were taken from different locations in the Jedeb watershed in South Gojjam at two depths of 0–15 cm and 15–30 cm. Soil water retention data was measured at matric potentials of six points (pF0, pF2.3, pF2.5, pF3, pF3.7, pF4.2) in Mekelle University Soil Physics laboratory. This data was not sufficient for calibration and validation of PTFs. In order to represent the soil heterogeneity present across the study area, it was essential to include water retention data from previous projects such as Upper Beles and Lake Tana projects (Table 35). It was very difficult to find more than four measured water content-matric potential (θ-h) pairs in the study area. Eventually only 123 water retention curves were found and used as input for subsequent analysis.

Data set 2 was compiled from seven sub-basin projects. *Data set 2* contains water contents at field capacity and permanent wilting point, PSD, BD, and OM (OM=Organic Carbon*1.72). *Data set 2* was used to evaluate effectiveness of Artificial Neural Network (ANN) and k-nearest neighbour (k-NN) point PTFs for the estimation of field capacity and permanent wilting point. A set of 614 samples was obtained after preliminary data screening and outlier removal and the summary statistics is given in Table 36.

126

In Table 34 *data set 3* is the result of a reconnaissance soil survey conducted as an integral part of the Abay River Basin Integrated Development Master Plan Project (BCEOM, 1998b). It was used to estimate and map the mean profile Available Water Content (AWC) based on the Soil Mapping Units (SMU) of the soil map of the Upper Blue Nile/Abay basin developed by BCEOM (1998b) at a scale of 1: 250, 000. A total of 53 SMUs was identified and each mapping unit is characterised by major landform and major soils. The particle size distribution or textural class and the depth of 890 horizons in each of the SMUs were collected.

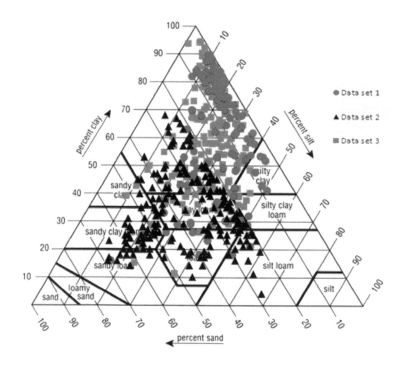

Figure 35 Textural distribution of the samples in the USDA textural triangle

Table 35 Summary statistics of the training data and test data for *data set 1***

	Sand	Silt	Clay	BD	OC	pH	CEC
	----------g g^{-1}--------			g cm^{-3}	%	----	cmol/kg
Training data (n=86)							
Minimum	0.52	8.00	18.00	0.50	0.10	4.20	2.15
Maximum	56.49	52.00	89.04	1.60	4.90	8.37	85.42
Mean	14.13	26.74	59.13	1.19	1.62	5.63	36.25
Standard Deviation	14.23	12.14	21.19	0.21	0.92	1.15	25.65
Test data (n=37)							
Minimum	0.64	8.18	16.00	0.93	0.26	4.20	3.36
Maximum	68.00	56.00	89.83	1.60	4.28	8.10	81.32
Mean	15.09	26.33	58.58	1.23	2.00	5.31	32.85
Standard Deviation	15.84	11.54	22.21	0.15	1.07	0.91	22.29

*BD = bulk density; OC= Organic Carbon; CEC = Cation Exchange Capacity

Table 36 Summary statistics of the training data and test data for *data set 2***

	Sand	Silt	Clay	BD	OM	$\theta 33$	$\theta 1500$
	----------g g^{-1}--------			g cm^{-3}	%	----cm^3 cm^{-3}---	
Training data (n=492)							
Minimum	0.00	0.05	0.04	0.50	0.15	0.15	0.06
Maximum	0.69	0.68	0.90	1.79	4.29	0.59	0.43
Mean	0.19	0.28	0.53	1.29	1.92	0.39	0.26
Standard Deviation	0.15	0.14	0.21	0.17	0.78	0.08	0.07
Test data (n=122)							
Minimum	0.01	0.04	0.14	0.60	0.48	0.18	0.10
Maximum	0.66	0.66	0.90	1.77	4.18	0.58	0.42
Mean	0.21	0.26	0.53	1.29	2.02	0.38	0.25
Standard Deviation	0.14	0.13	0.21	0.18	0.81	0.08	0.07

*BD = bulk density; OM = organic matter; $\theta 33$ = water retention at -33 kPa matric potential; $\theta 1500$ = water retention at -1500 kPa matric potential

8.3.2   Derivation of Pedo-transfer Functions

Parameterization

Measuring a soil water retention curve ($\theta(h)$) is laborious, expensive and time consuming. Measured values of soil water retention curves are often fragmentary, and usually constitute relatively few θ-h pairs. For modeling, characterization and comparison purposes, it is therefore important to describe the whole of the soil water retention curve in a parametric form. In this study, the unsaturated soil hydraulic functions were described using a commonly used parametric model called the van Genuchten (VG) model (van Genuchten, 1980):

$$\theta(h) = \begin{cases} \theta_r + \frac{\theta_s - \theta_r}{(1+|\alpha h|^n)^m}, & h < 0 \\ \theta_s & h \geq 0 \end{cases} \qquad [8.1]$$

where $\theta(h)$ is the volumetric water content at potential h (cm); θ_r is the residual water content (cm^{3}cm^{-3}) i.e. the water content at which the gradient $d\theta/dh$ becomes zero when h becomes largely negative; θ_s is the saturated water content (cm^{3}cm^{-3}); α, n, and m are the shape parameters (-).

The retention data of *data set 1* was fitted to Eq. 8.1. Based on the measured and available data the following water potentials were used to optimize the VG model parameters: pF 0, 2.3, 2.5, 3.0, 3.5, 3.7 and 4.2. In the optimization process θ_s was fixed to the measured saturated water content (when available) and Mualem's restriction of $m = 1-(1/n)$, with n>0, was implemented in order to reduce the number of model parameters to be optimized. The remaining parameters (θ_r, α, and n) were optimized by fitting Eq. 1 to the measured θ-h data by means of a weighted non-linear least square approach based on the Marquardt's method, as provided by the RETC (RETention Curve) program (Van Genuchten et al., 1991). However, during the fitting process most of the measured water retention data resulted in fitted values for θ_r that were set automatically to zero. Therefore, θ_r was fixed at the value of zero throughout.

Deriving class pedotransfer functions

Four steps were implemented in deriving class PTFs using the *Wösten method* (Wösten et al., 1995): First, the optimized VG model parameters were used to determine the water contents at 14 matric potentials: 0, 1.0, 1.3, 1.7, 2.0, 2.3, 2.5, 2.7, 3.0, 3.5, 3.7, 4.0, and 4.2 pF. Second, the derived water retention curves were primarily divided into topsoil (A horizons) and subsoil (B and C horizons). Third, the groups were further subdivided by FAO/USDA textural classes to compute the geometric mean water contents at the selected matric potentials. The geometric mean is preferable to the arithmetic mean for log-normally distributed data. Besides, the water content values within one standard deviation were calculated in order to determine the degree of variation of individual curves around the geometric mean. Finally, the 14 geometric mean values were again fitted to Eq. 1 to get the VG parameters of the $\theta(h)$ relationships.

Deriving continuous and point pedotransfer functions

Multiple regression techniques
The dependency of the VG parameters (θ_r, θ_s, α, and n) on the more easily measured basic soil properties was investigated using multiple regression techniques to derive parametric (continuous) PTFs. In developing point PTFs, water retention at specific

matric potentials of 0, 2.3, 2.5, 3.0, 3.5, 3.7, and 4.2 pF were related to basic soil properties using Eq. 8.2. Linear, reciprocal, quadratic and logarithmic relationships of soil properties along with interactions were used in the initial step of the regression analysis. The general form of the regression equation is given by

$$Y = \beta_0 + \beta_1 Sa + \beta_2 Si + \beta_3 Cl + \beta_4 OC + \beta_5 CEC + \beta_6 pH + \beta_7 BD$$

$$+\beta_8 Sa^{-1} + \cdots + \beta_{14} BD^{-1} + \beta_{15} Ln(Sa) + \cdots + \beta_{21} Ln(BD)$$

$$+\beta_{22} Sa^2 + \cdots + \beta_{28} BD^2 + \beta_{29} Sa * Si + \cdots + \beta_n pH * BD \qquad [8.2]$$

where Y refers to θ_s or $\ln(\alpha)$ or $\ln(n\text{-}1)$, β_0 is the intercept, and $\beta_{1, ..., } \beta_n$ are regression coefficients.

The selection of potential subsets of predictor variables can be performed by the *all possible subsets, stepwise forward selection,* and *stepwise backward elimination* methods (Sheather, 2009). However, the latter two methods are criticized for fewer combinations of sub-models and rarely select the best one (Hocking, 1976; Sheather, 2009). In this study, the *all possible subset* method was employed. This method performs an exhaustive search for the best subsets of potential predictive variables, based on the leaps-and-bounds algorithm of Furnival and Wilson (1974). If there are p predictors, the *all possible subset* method selects $p + 1$ most optimal models having the smallest residual sum of square from a total of 2^p possible models. A single best model was selected among the $p+1$ options based on the following criteria (Izenman, 2008;Sheather, 2009): a model with the largest value of adjusted R^2; a model with the smallest value of Bayes information criterion (BIC); a model with the smallest value of Akaike information criterion (AIC); and a model with the smallest Mallows's Cp value (Hocking, 1976). AIC and Cp are proportional to each other (Izenman, 2008), and thus only Cp was used. This procedure was carried out by the *leaps* function (*regsubsets*) that is implemented in R software. The final best single model was selected after checking for violations of any regression assumptions. Back-transformation to the original VG model parameter was necessary to obtain the hydraulic characteristics.

Once the proposed continuous PTFs are developed and tested, comparisons of the performance of the proposed PTFs and that of the continuous tropical PTFs of Hodnett and Tomasella (2002) were carried out. The PTFs of Hodnett and Tomasella (2002) were derived from 771 tropical soil horizons of the IGBP-DIS soil database using multiple regression techniques. Soils of the world between 25°N and 25°S were included in the PTF development. However, tropical soils found in temperate climates due to high altitude were not considered by the PTFs of Hodnett and Tomasella (2002).

Data mining and exploration techniques for point PTFs
In this study, point PTFs were developed and tested using both ANN and *k*-NN techniques. A three-layer feed-forward backpropagation ANN model with multilayer perceptron (MLP) was used to develop ANN based PTFs. The input layer had three to five neurons that received the percentage sand, silt and clay, bulk density, and

organic matter as input hierarchically. The number of neurons in the hidden layer was determined by trial and error; three hidden nodes gave the best results for the data used in this study. The signal in the output layer represented the corresponding moisture content at field capacity and permanent wilting point. The Levenberg-Marquardt training algorithm with the *logistic* activation function in the hidden layer and a *hyperbolic tangent* activation function in the output layer were used. In this study, ANN model development was performed by using the NeuroSolutions 5 software. The other alternative approach to statistical regression is the *k*-NN method (Jagtap et al., 2004; Nemes et al., 2006). This technique is based on pattern recognition and similarity used to identify and retrieve the most similar or nearest stored objects to the target object. In this study, software developed by Nemes et al. (2008) was used in order to build *k*-NN PTFs. *Data set 2* was split into a training (OECD) set with 492 samples and a test set including 122 samples. Four levels of *k*-NN and ANN PTFs were developed relating soil water retention at -33 kPa and -1500 kPa to the following basic soil properties:

(a) sand, silt, clay (SSC)
(b) sand, silt, clay, and bulk density (SSCBD)
(c) sand, silt, clay, and organic matter (SSCOM)
(d) sand, silt, clay, bulk density and organic matter (SSCBDOM).

Analysis of variance was run to determine whether differences between ANN and *k*-NN models in predicting soil water contents at field capacity and permanent wilting point are significant or not.

8.3.3   Mapping Potential Available Water Capacity

The derived class PTFs were used to compute the mean AWC for Soil Mapping Units (SMU) of the Upper Blue Nile/Abay basin. Available water was considered as the water held between wilting point (pF=4.2) and field capacity (pF=2.5). First, the depth and texture of topsoil and subsoil for different soil profiles within each SMU were obtained from the available descriptions of SMU attributes (BCEOM, 1998b). Second, the water contents at field capacity (θ_{FC}) and wilting point (θ_{WP}) were determined from the class PTFs using Table 37. Next, the available water for each topsoil and subsoil of the available soil profiles in each SMU was computed by multiplying the water content held between field capacity and wilting point (θ_{FC}- θ_{WP}) with the thickness of each horizon. Finally, the total Available Water Capacity (AWC) for each profile was calculated by summing up the available water of topsoil and available water of subsoil.

8.3.4   Model Performance Evaluation

The coefficient of determination (R^2), Mean Residual, Root Mean Square Residual (RMSR), and Unbiased Root Mean Square Residual (URMSR) were used to compare and evaluate predicted and measured water contents at specific water potentials. R^2 indicates how much of the variability can be explained by its relationship to another factor. It is computed as

$$R^2 = 1 - \frac{\sum_{i=1}^{N}(\theta_{m,i}-\theta_{p,i})^2}{\sum_{i=1}^{N}(\theta_{m,i}-\bar{\theta})^2} \qquad [8.3]$$

where $\theta_{m,i}$ are measured values, $\theta_{p,i}$ are predicted values at time/place i and $\bar{\theta}$ is the average value of the measured data.

MR quantifies the systematic errors (bias) between observed and modeled data. One of the advantages of using MR is that it indicates whether a PTF over-estimates (MR > 0) or under-estimates (MR < 0) the water content.

$$MR = \frac{1}{N}\sum_{i=1}^{N}(\theta_{p,i} - \theta_{m,i}) \qquad [8.4]$$

RMSR can be interpreted as the standard deviation of the unexplained variance. In other words it provides information about the accuracy of predictions in terms of standard deviation.

$$RMSR = \sqrt{\frac{\sum_{i=1}^{N}(\theta_{p,i}-\theta_{m,i})^2}{N}} \qquad [8.5]$$

If mean errors are present then the RMSR given by Eq. 8.3 includes both systematic and random errors. URMSR was also computed in order to measure the precision of prediction alone by separating the systematic from the random errors (Tietje and Tapkenhinrichs, 1993). URMSR is given by

$$URMSR = \sqrt{\frac{\sum_{i=1}^{N}(\theta_{p,i}-\theta_{m,i}-MR)^2}{N}} \qquad [8.6]$$

Mean Difference (MD) and Root Mean Square Difference (RMSD) were computed using MATLAB software to evaluate continuous soil water retention curves (SWRC) by integrating the measured and predicted functions between top and bottom boundaries (Tietje and Tapkenhinrichs, 1993). Cornelis et al. (2001) pointed out that integrating the measured and the predicted functions between lower and upper boundaries makes the evaluation more objective than simple summation. Lower and upper boundary were set to log (1 cm) and log (15,849 cm), respectively, both are within the measured water retention data.

The MD (cm^3cm^{-3}) for soil sample i was computed as

$$MD_i = \frac{1}{b-a}\int_a^b \left[\theta(h)_{p_i} - \theta(h)_{m_i}\right] dh \qquad [8.7]$$

The RMSD (cm^3cm^{-3}) was computed as

$$RMSD_i = \sqrt{\frac{1}{b-a}\int_a^b \left[\theta(h)_{p_i} - \theta(h)_{m_i}\right]^2 dh} \qquad [8.8]$$

where $\theta(h)_{m_i}$ is the measured soil water retention function for soil sample i (continuous VG curve fitted to the measured discrete $\theta(h)$ pair data), $\theta(h)_{p_i}$ is the predicted soil water retention curve for soil sample i as predicted by a PTF, a and b are the lower and upper integration boundaries, respectively.

8.4 RESULTS AND DISCUSSION

8.4.1 The derived class pedotransfer functions

Table 37 represents the typical values of the VG model parameters for each textural class resulting from the fitting of Eq. 1 to the geometric mean water contents at 14 matric potentials. The R^2 and SSE values indicate that the calculated geometric means and the fitted curves agree very well. Figure 36 shows the calculated geometric mean water retention and the standard deviations for very fine texture of topsoil and subsoil.

Table 37 VG model parameters for the fits on the geometric mean curves

Horizon	Texture	θ_r	θ_s*	α	n	R^2	SSE
		FAO texture classes					
Topsoil	Very fine	0.0000	0.5390	0.0104	1.3168	0.9958	0.00100
	Fine	0.0000	0.4900	0.0326	1.1691	0.9984	0.00020
	Medium	0.0000	0.5838	0.0327	1.1948	0.9968	0.00080
Subsoil	Very fine	0.0000	0.5192	0.0166	1.2838	0.9983	0.00050
	Fine	0.0000	0.5027	0.0346	1.1809	0.9966	0.00060
	Medium	0.0000	0.5448	0.0164	1.2058	0.9996	0.00010
		USDA texture classes					
Topsoil	Clay	0.0000	0.5124	0.0207	1.1984	0.9979	0.00040
	Clay loam	0.0000	0.5500	0.0189	1.2037	0.9944	0.00130
	Loam	0.0294	0.5191	0.0333	1.1668	0.9998	0.00003
	Silty clay	0.0138	0.5012	0.0149	1.1873	0.9997	0.00004
Subsoil	Clay	0.0000	0.5077	0.0218	1.2480	0.9973	0.00070
	Clay loam	0.0000	0.5703	0.0261	1.1871	0.9987	0.00031
	Loam	0.0000	0.5906	0.0230	1.1726	0.9991	0.00021
	Silty clay	0.0286	0.5045	0.0066	1.2738	0.9998	0.00004

*Parameter fixed at measured value

Medium textured subsoil showed decreased water retention at matric suction less than 20 cm as compared to medium textured topsoil, whereas similar water retention characteristics were observed at all matric suctions greater than 20 cm (Fig. 36a). The difference in compaction between topsoil and subsoil horizon of medium textured soil resulted in a pronounced effect in the low-suction range of the SWRC. In the very fine textured soil, greater water retention of topsoil was observed at any particular suction (Fig. 36a&b). This is due to the fact that the topsoil has greater clay content and organic matter content than that of the subsoil. At higher suctions the difference between the topsoil and subsoil SWRCs of very fine textured soil is significant, while at lower suctions this difference is very small. This suggests that the textural attribute of soil water retention of very fine textured soil is stronger than its structure attribute. These observations clearly indicate that the structure can markedly affect the soil water retention of medium textured soil, whereas the effect of soil structure is less pronounced in very fine textured soil.

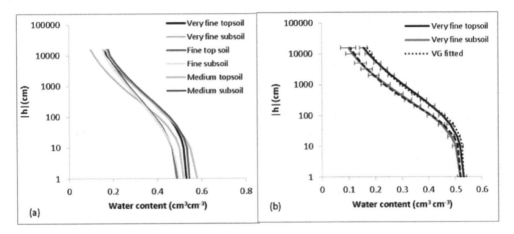

Figure 36 (a) Computed geometric mean water retention characteristics for three FAO textural classes (very fine, fine and medium) for both topsoil and subsoil, (b) computed geometric mean water retention characteristics (solid lines) and the standard deviations for very fine textured soil (bars). The van Genuchten fit to the geometric mean values are shown as a dotted line

Application of class PTFs to Upper Blue Nile/Abay basin

The error bars in Fig. 36b represent the variations in individual water retention characteristics used to calculate the average characteristics. Since the individual water retention properties within each class depict considerable variability, the accuracy of class PTF is limited. However, class PTFs are easy to use, require little data input, and are well suited for predicting water retention properties over large areas. This was demonstrated by the use of class PTFs to map the available water content of the whole Upper Blue Nile/Abbay basin (Fig. 37). The eastern part of the basin which is mountainous and has shallow soil showed lower available water content than the western part of the basin (Fig. 34). The mean available water content of the basin is 144.5 mm. Leptosols which represent the most widely occurring soils within the basin appear to have a low AWC of 96.9 mm. The soils with highest available water content are Fluvisols (mean of 188.5 mm) and Acrisols (mean of 183.1 mm). Overlay analysis of land use/land cover map of the basin on the soil map revealed that the majority of the wetlands were found to be located on soils of highest available water content (i.e. Fluvisols).

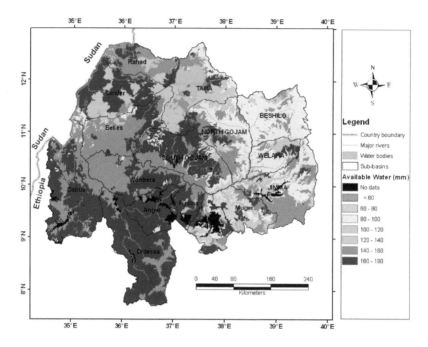

Figure 37 Map of the potential available water contents ($AWC_{potential} = (\theta_{FC} - \theta_{WP})$*depth) of Upper Blue Nile/Abbay basin soil profiles of variable depth derived by using class PTFs; see Table 38 for input data

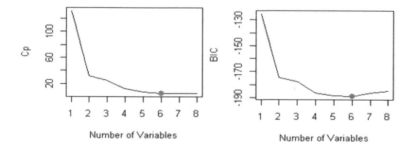

Figure 38 Mallow's Cp and Bayesian Information Criterion (BIC) are displayed for the best models of each size for the saturated water content (θ_s)data set

8.4.2   The Derived Continuous Pedotransfer Functions

The optimal regression models

Models used to estimate each parameter were selected based on the following criteria: the largest value of adjusted R^2, the smallest value of Bayes information criterion (BIC), the smallest Mallows' Cp value. For instance, Fig. 38 & 39 indicate that the model with the largest adjusted R^2, lowest Cp and lowest BIC values was the six-predictor model for estimating θ_s. Similarly, the five-predictor model and the four-predictor model were selected for estimating $\ln(\alpha)$ and $\ln(n-1)$, respectively. The response variables $\ln(\alpha)$ and $\ln(n-1)$ were de-transformed (inversed) to refer to the original VG model parameters (i.e. α and n). The selected optimal models for estimation of the parameters θ_s, α, and n are presented in Table 38.

Table 38 Selected optimal regression models. Basic soil properties such as percent sand (Sa), percent clay (Cl), percent silt (Si), percent Organic Carbon (OC), Cation Exchange Capacity (Dinku et al.), pH, and bulk density (BD) were used as independent variables.

Parameters	PTFs
	Parametric PTFS
θ_s (cm^{3}cm^{-3})	$0.976 - 0.497BD - 0.0043OC^{-1} + 3.04Cl^{-1} + 0.00059CEC*BD + 0.001Cl*BD - 0.135CEC^{-1}$
$\alpha*$(cm^{-1})	$-3.29 - 0.727Ln(Sa) - 0.227pH*BD - 0.0153CEC*BD + 0.003Sa*Cl + 0.0008Si*Cl$
$n*$(-)	$-1.46 + 0.011CEC - 0.019\ Sa*BD + 5.56*10^{-4}Sa*Si - 3.02*10^{-4}Si*Cl$
	Point PTFs
θ_{200cm} (cm^{3}cm^{-3})	$-0.141 + 2.8Cl^{-1} + 0.446\ BD^{-1} + 4.44*10^{-3}CEC*BD$
θ_{330cm} (cm^{3}cm^{-3})	$0.39 + 0.018\ Ln(OC) + 0.014\ Ln(Sa) + 1.94*10^{-3}pH^2 - 0.059\ BD^2 - 7.7*10^{-5}\ Sa*Cl$
θ_{1000cm} (cm^{3}cm^{-3})	$0.298 - 0.227\ pH^{-1} + 0.681Si^{-1} + 4.74Cl^{-1} - 0.024BD^2 - 5.6*10^{-5}\ Sa*Cl$
θ_{3000cm} (cm^{3}cm^{-3})	$0.365 - 4.86\ CEC^{-1} - 0.0090C^{-1} + 0.321Si^{-1}$
θ_{5000cm} (cm^{3}cm^{-3})	$0.075 + 0.4\ pH^{-1} + 0.523*Si^{-1} + 0.034Ln(Sa) + 0.7*10^{-5}CEC^2 - 0.6*10^{-5}Sa^2$ $- 6.9*10^{-5}Sa*Cl$
$\theta_{15000cm}$ (cm^{3}cm^{-3})	$0.104 - 0.1312Sa^{-1} + 0.008Ln(Dinku\ et\ al.) + 5.2*10^{-5}Sa^2 - 0.071BD^2 + 0.014pH*BD$ $+ 0.0014Cl*BD$

$\alpha* = \ln(\alpha)$, $n* = \ln(n-1)$,

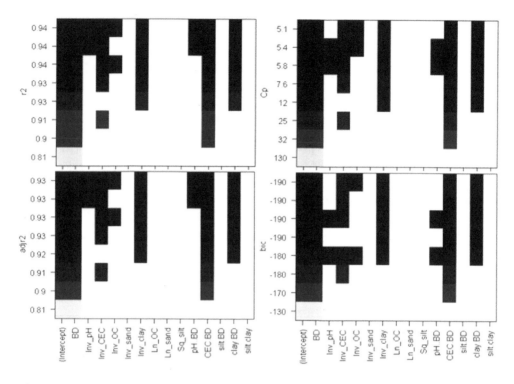

Figure 39 Graphical presentation of best subset models for predicting θs. Each row in this graph represents a model; the shaded rectangles in the columns indicate the variables included in the given model; the numbers on the left margin are the values; the darkness of the shading simply represents the ordering of the BIC values

A Pearson correlation analysis of soil hydraulic properties versus basic soil properties exhibited a statistically significant negative correlation of saturated water content (θ_s) with bulk density ($r = -0.85$, $p = 0.000$) and soil pH ($r = -0.22$, $p = 0.013$), while positive correlation with organic carbon content ($r = 0.27$, $p = 0.002$). The PTF for θ_s given in Table 38 also revealed these relationships. Values of α and the clay content are negatively correlated indicating that the larger the amount of clay in the soil, the smaller α becomes. This means that there is little change of water content at higher suctions. The PTF of $\ln(\alpha)$ in Table 38 also showed that the parameter α is determined by soil structure (bulk density) as well. The parameter n showed a statistically significant positive correlation with soil pH ($r = 0.24$, $p = 0.008$), CEC ($r = 0.37$, $p = 0.000$), and clay content ($r = 0.27$, $p = 0.003$) and a negative correlation with sand content ($r = -0.26$, $p = 0.003$). Therefore, these evidences suggest that soil water retention depends not only on soil structure (i.e. bulk density and organic carbon) or soil texture, but also on its chemical properties such as soil pH and CEC. Pachepsky and Rawls (1999) also found similar results.

Table 39 Performance of the PTFs developed with *all subset* regression analysis

Parameters	Calibration data set			Testing data set		
	R^2	RMSR	URMSR	R^2	RMSR	URMSR
			Continuous PTFs			
$\theta_s(cm^3cm^{-3})$	0.94	0.024	0.023	0.93	0.019	0.018
$\alpha*(cm^{-1})$	0.51	1.109	1.108	0.51	1.034	1.032
$n*(-)$	0.57	0.270	0.270	0.57	0.297	0.294
			Point PTFs			
$\theta_{200cm}(cm^3cm^{-3})$	0.93	0.022	0.022	0.93	0.047	0.043
$\theta_{330cm}(cm^3cm^{-3})$	0.62	0.032	0.032	0.59	0.042	0.042
$\theta_{1000cm}(cm^3cm^{-3})$	0.53	0.026	0.026	0.51	0.044	0.038
$\theta_{3000cm}(cm^3cm^{-3})$	0.51	0.031	0.031	0.49	0.030	0.029
$\theta_{5000cm}(cm^3cm^{-3})$	0.43	0.021	0.021	0.40	0.019	0.019
$\theta_{15000cm}(cm^3cm^{-3})$	0.61	0.017	0.017	0.60	0.022	0.018

$\alpha* = \ln(\alpha)$, $n* = \ln(n-1)$

The model for predicting θ_s explained 93% of the variability using predictors such as clay content, bulk density and CEC. However, substantial unexplained variability was observed in the $\ln(\alpha)$ (51% explained variability) and $\ln(n-1)$ (57% explained variability) PTFs (Table 39). Use of large database and incorporation of additional predictors such as clay mineralogy and categorical soil structure information could increase the explained variability in the prediction of α and n parameters. The prediction of θ_s was relatively better as indicated by lower RMSR both in the calibration and testing data sets, whereas the performance of the PTF for the parameter $\ln(\alpha)$ was less accurate. Estimations were also made for soils that are outside the range of soils used for developing the PTFs, as the performance of PTFs is largely dependent on the data used in deriving the PTFs (Schaap et al., 1998;Tombul et al., 2004). In general, the application of continuous PTFs showed a good agreement between observed and predicted values in both calibration and testing data sets (Table 40).

Comparison with existing tropical PTF

Table 40 presents the performance evaluation statistics of the derived PTFs and the PTF of Hodnett and Tomasella (2002) for both calibration and test data sets. The smaller values of the means of the absolute values of MD (MAMD) and of RMSD (MRMSD) indicate better performance of the PTF. The MAMD of the proposed PTF (0.0134 cm^3cm^{-3}) is smaller than that of PTF of Hodnett and Tomasella (2002) (0.0787 cm^3cm^{-3}) indicating lower prediction error of the proposed PTF. The mean of root mean square difference (MRMSD) of the proposed PTF was more than twice as low as MRMSD of PTF of Hodnett and Tomasella (2002) with values of 0.0508 and 0.0996 cm^3cm^{-3}, respectively. Thus, the soil water retention curve (SWRC) predicted by the proposed PTF follows the shape of the measured SWRC better than the PTF of Hodnett and Tomasella (2002) does for soils of the Upper Blue Nile/Abay area.

Validation was carried out using an independent testing data set, which had not been used in the derivation of the PTFs. The model fit for the validation data set and the calibration data set is almost similar. Therefore, developing the proposed PTFs offered a considerable importance for proper characterization of SWRC of high altitude tropical soils. Tietje and Tapkenhinrichs (1993) found out that MRMSD ranges from 0.04 to 0.13 cm^3cm^{-3} after applying several PTFs to many soils and the MRMSD values of both PTFs fall within this range of values. When considering standard deviation of RMSD, it can be observed that the proposed PTF performs better than that of Hodnett and Tomasella (2002) (0.0434 and 0.0606 cm^3cm^{-3}, respectively). This again indicates that the performance of the proposed PTF is relatively less dependent on the soil type than the PTF of Hodnett and Tomasella (2002). Regarding the mean of MDs (MMD), the proposed PTF tends to underestimate while the PTF of Hodnett and Tomasella (2002) tends to overestimate in the calibration data set. In the testing data set both PTFs show the tendency to overestimate. A good agreement was observed between measured and predicted water contents in both calibration and testing data sets of both point and continuous PTFs (Table 39 & 40).

Table 40 Prediction performance of the derived PTFs and existing tropical PTFs

PTFs	Calibration data set				Testing data set			
	MMD	MAMD	MRMSD	SD RMSD	MMD	MAMD	MRMSD	SD RMSD
Proposed Point PTFs	0.0592	0.0642	0.0378	0.0347	0.0658	0.0675	0.0349	0.0181
Proposed continuous PTFs	-0.0023	0.0318	0.0421	0.0347	0.0134	0.0361	0.0508	0.0434
Hodnett and Tomasella (2002)	0.0801	0.0839	0.0990	0.0543	0.0787	0.0787	0.0996	0.0454

Figure 40 depicts that the PTF of Hodnett and Tomasella (2002) predicts better at the very wet end (near saturation) of the SWRC (i.e. pF = 0) than that of the proposed PTF. However, better prediction by the proposed PTF is observed starting from a pF value of 3 up to the dry end (pF = 4.2) of the retention curve. Water retention at the wet end (low matric suction) is strongly affected by the structure of the soil (Hillel, 2004) and structure can markedly affect the water retention of medium textured soil (Fig. 40a). Therefore, including more samples from medium textured soil could improve the predictive capacity of the proposed PTF at the wet end.

Tropical soils developed under temperate climate due to high altitude were not considered in the derivation of the PTF of Hodnett and Tomasella (2002). In contrast, the proposed PTF was derived mainly using soils within the tropics but in temperate climate because of the altitude. This could be one of the reasons for the poor fit at the dry end of the soil water retention curve for the PTF of Hodnett and Tomasella (2002).

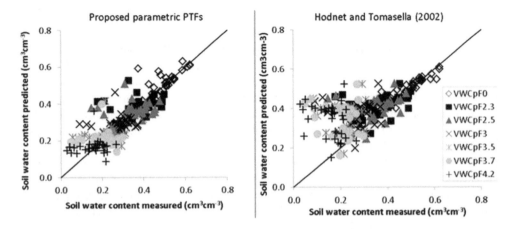

Figure 40 Volumetric water content (VWC) as predicted by the proposed PTF and the PTFs of Hodnett and Tomasella (2002)

8.4.3   The Derived Point Pedotransfer Functions

Regression based point PTFs

Table 38 depicts the derived point PTFs predicted water contents at matric potentials of -200, -330, -1000, -3000, -5000, -15,000 cm. The predictions of water contents at all matric potentials for the calibration data set and independent testing data set resulted in a closely related model fit, as indicated by similar R^2 values in both data sets. The precision of prediction of the model at -15,000 cm was the best as indicated by the URMSR value of 0.018 in the testing data set, whereas the precision of the model at -200 cm is the least (URMSR=0.047). In general, the precision of prediction increases in the dry range of the SWRC. Indices of soil structure such as bulk density tend to strongly affect the water retention properties at water potentials of 0 and -200 cm. Bulk density also influenced the water retention at -330, -1000, and -15000 cm but its effect was weak. Organic carbon content affected the water retention at -330 and -3000 cm, but not at -15,000 cm and others. Calhoun et al. (1973) also found a similar result that the use of organic carbon content improved prediction of water contents at -330 cm, but not at -15,000 cm. In contrast, Beke and MacCormick (1985) showed the importance of organic carbon in the prediction of water contents at -15,000 cm, but not at -330 cm. Either bulk density or the organic carbon content tends to influence water retention together with the soil chemical (pH and CEC) and the soil texture properties.

ANN and k-NN-based point PTFs

Table 41 depicts the performance evaluation statistics of PTFs developed using k-NN and ANN methods. In estimating soil water retention at -330 cm (θ_{FC}), the performance of the k-NN PTF was relatively better in all hierarchical models as indicated by lower values of MR, URMSR, RMSR and higher R^2. However, the performance of the ANN PTF was better at estimating the water retention at -15000 cm (θ_{PWP}) in most of the input levels except for SSCOM. The addition of bulk density to the percent sand, silt, and clay did not improve the accuracy of both the k-NN PTF and ANN PTF in estimating θ_{FC}. However, the addition of bulk density markedly improved the accuracy of both k-NN and ANN PTF in predicting θ_{PWP} as indicated by the decrease in RMSE from 0.0585 to 0.0570 cm^{3}cm^{-3} in ANN PTFs and from 0.0597 to 0.0585 cm^{3}cm^{-3} in k-NN PTFs. In general, in predicting θ_{FC} k-NN PTFs had lower mean RMSR (0.0643 cm^{3}cm^{-3}) as compared to ANN PTFs (0.0670 cm^{3}cm^{-3}). However, in predicting θ_{PWP} ANN PTFs had lower mean RMSR (0.0578 cm^{3}cm^{-3}) as compared to k-NN PTFs (0.0585 cm^{3}cm^{-3}). The addition of organic matter content to percent sand, silt and clay markedly increased the accuracy of k-NN PTFs in estimating both θ_{FC} and θ_{PWP} (Fig. 41a-d) as the RSMR value decreased from 0.0644 to 0.0631 cm^{3}cm^{-3} in θ_{FC} and from 0.0597 to 0.0576 cm^{3}cm^{-3} in θ_{PWP}. However, the inclusion of organic matter did not improve the performance of ANN PTFs as indicated by the increase in URMSR from 0.0684 in SSC to 0.0687 in SSCOM in estimating θ_{FC} and an increase in URMSE from 0.0621 to 0.0624 in estimating θ_{PWP}. Therefore, there is no need to include additional soil parameters such as organic matter, as with sole texture data a better accuracy can be achieved.

With ANN PTFs, the greatest prediction bias (MR=0.0118 cm^{3}cm^{-3}) and the least precision (URMSR=0.0687 cm^{3}cm^{-3}) were observed in the SSCOM model in the estimation of both θ_{FC} and θ_{PWP}. With k-NN PTFs, the greatest bias is observed again in the SSCOM model, while the least precision was recorded in the SSC model. In general, greater prediction bias but higher precision was involved in the estimation of θ_{PWP} as compared to that of θ_{FC} for both ANN and k-NN PTFs.

Table 41 Performance evaluation statistics for the hierarchical ANN and k-NN-based
point PTFs

Response variable	Models	ANN				k-NN			
		R^2	MR	URMSR	RMSR	R^2	MR	URMSR	RMSR
θ_{FC}	SSC	0.33	0.0081	0.0684	0.0672	0.38	0.0052	0.0644	0.0644
	SSCBD	0.33	0.0063	0.0686	0.0669	0.38	0.0086	0.0640	0.0644
	SSCOM	0.32	0.0088	0.0687	0.0672	0.41	0.0088	0.0628	0.0631
	SSCBDOM	0.34	0.0066	0.0677	0.0666	0.36	0.0076	0.0651	0.0653
	Mean	*0.33*	*0.0075*	*0.0684*	*0.0670*	*0.38*	*0.0076*	*0.0641*	*0.0643*
θ_{PWP}	SSC	0.34	0.0116	0.0621	0.0585	0.32	0.0116	0.0588	0.0597
	SSCBD	0.36	0.0093	0.0594	0.0570	0.34	0.0119	0.0575	0.0585
	SSCOM	0.33	0.0118	0.0624	0.0585	0.37	0.0135	0.0563	0.0576
	SSCBDOM	0.36	0.0093	0.0603	0.0571	0.35	0.0120	0.0573	0.0583
	Mean	*0.35*	*0.0105*	*0.0611*	*0.0578*	*0.35*	*0.0123*	*0.0575*	*0.0585*

There was a statistically significant interaction between the effects of methods used
and the type of response variable (F (1, 12) = 10.52, P = 0.007) (Table 42). There
was also a statistically significant interaction between the effects of methods used and
the type of response variable on RMSR (F (1, 12) = 19.82, P = 0.001). In other
words, the differences in R^2 and RMSR between the ANN and k-NN methods are
dependent on the type of the predicted parameters (FC or PWP). The mean RMSR
of k-NN is significantly lower than that of ANN in predicting FC, whereas the two
methods are not statistically different in predicting PWP in terms of RMSR.

Table 42 ANOVA results showing the effect of the two methods (ANN and k-NN) and type
of the response variable (FC and PWP) on performance statistics (i.e. the coefficient of
determination (R^2), Mean Residual (Sikka et al.), Root Mean Square Residual (RMSR), and
Unbiased Root Mean Square Residual (URMSR))

Performance evaluation statistics	Method		Response variable		Interaction	
	F	P	F	P	F	P
R^2	1.391	0.261	8.696	0.012*	10.522	0.007*
MR	2.010	0.182	35.279	0.000*	1.599	0.230
URMSR	57.884	0.000*	181.489	0.000*	0.460	0.510
RMSR	6.261	0.028*	378.882	0.000*	19.819	0.001*

* Significant at $p < 0.05$

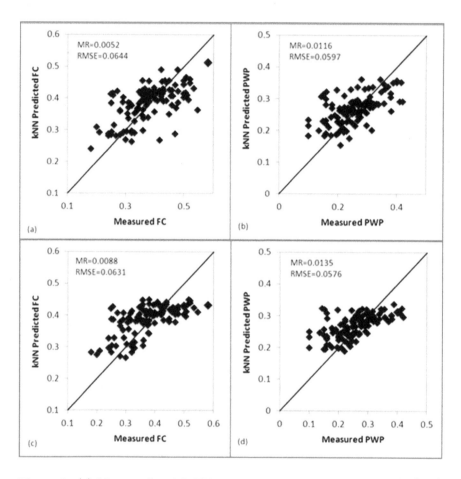

Figure 41 (a) Measured and k-NN predicted soil water retention (cm³cm⁻³) at FC (-33 kPa) using SSC (b) Measured and k-NN predicted soil water retention (cm³cm⁻³) at PWP (-1500 kPa) using SSC (c) Measured and k-NN predicted soil water retention (cm³cm⁻³) at FC (-33 kPa) using SSCOM (d) Measured and k-NN predicted soil water retention (cm³cm⁻³) at PWP (-1500 kPa) using SSCOM

8.5 CONCLUSIONS

Point and continuous PTFs with the largest adjusted R^2, lowest Cp and lowest BIC values were developed using the *all possible subset* regression method. The results showed that a good agreement between observed and predicted values in both calibration and testing data sets could be achieved. The PTFs revealed that soil water retention depends, not only on soil structure (i.e. bulk density and organic carbon) or soil texture but also on its chemical properties such as soil pH and CEC. The overall prediction performance of both the derived point and continuous PTFs were found to be superior to the tropical PTFs of Hodnett and Tomasella (2002). This could be explained by the fact that tropical soils developed under temperate climate due to high altitude were not considered in the derivation of the PTF of Hodnett and Tomasella (2002). In contrast, the proposed PTF was derived mainly using soils within the tropics in high altitude area. The prediction error of the proposed continuous PTFs increases with increasing sand content and decreasing bulk density. This suggests that future studies are recommended to be carried out on a database that is large enough to be representative of almost all parts of the textural triangle.

The derived regression-based point PTFs indicated that either bulk density or organic carbon content tend to influence water retention together with the soil chemical (pH and CEC) and the soil texture properties. Indices of soil structure such as bulk density tend to strongly affect the water retention properties at water potentials of 0 and -200 cm. The use of organic carbon content improved prediction of water contents at water potentials of -330 cm, but not at -15000 cm.

Analysis of four levels of ANN and k-NN-based point PTFs relating θ_{FC} and θ_{PWP} to different levels of basic soil data demonstrated the predictive ability achieved by adding parameters. The addition of bulk density to the percent sand, silt, and clay did not improve the accuracy of both the k-NN PTF and ANN PTF in estimating θ_{FC}. However, the addition of bulk density markedly improved the accuracy of both k-NN and ANN PTF in predicting θ_{PWP}. The addition of organic matter content to percent sand, silt and clay markedly increased the accuracy of k-NN PTFs in estimating both θ_{FC} and θ_{PWP}. However, the inclusion of organic matter did not improve the performance of ANN PTFs. Therefore, there is no need to include additional soil parameters such as organic matter, as with sole texture data a better accuracy can be achieved. The two-way ANOVA analysis revealed that difference in R^2 and RMSR between ANN and k-NN methods is dependent on the type of the predicted parameters (FC or PWP). The mean RMSR of k-NN is significantly lower than that of ANN in predicting FC, whereas the two methods are not statistically different in predicting PWP in terms of RMSR in the Upper Blue Nile area.

The difference in compaction between topsoil and subsoil horizon of medium textured soil resulted in a pronounced effect in the low-suction range of the SWRC. In contrast, in the very fine textured soil, the effect of compaction was pronounced at higher suctions. These suggested that structure can markedly affect the soil water retention of medium textured soil, whereas the effect of soil structure is less pronounced in very fine textured soils. With only texture class information, class PTFs enables users to estimate the water retention properties, hence class PTFs are cheap and easy to use. This was demonstrated by mapping the total available water capacity through the use of the derived class PTFs for the Upper Blue Nile/Abay basin using the 1:250,000 scale SMU of the available soil map. Thus, collecting and integrating fragmented available soil survey data enables scientists and planners to develop area-specific PTFs and generate new information such as maps of available water at the basin scale.

PART 3: PATTERNS AND CLIMATIC CONTROLS OF

VEGETATED LAND COVER DYNAMICS

Chapter 9

INTER-ANNUAL AND SEASONAL TRENDS OF VEGETATION CONDITIONS: DUAL SCALE TIME SERIES ANALYSIS[7]

9.1 ABSTRACT

A long-term decline in ecosystem functioning and productivity, often called land degradation, is a serious environmental and development challenge to Ethiopia that needs to be understood so as to develop sustainable land use strategies. This study examines inter-annual and seasonal trends of vegetation cover in the Upper Blue Nile (UBN) or Abbay basin. Advanced Very High Resolution Radiometer (AVHRR) based Global Inventory, Monitoring, and Modelling Studies (GIMMS) Normalized Difference Vegetation Index (NDVI) was used for long-term vegetation trend analysis at low spatial resolution. Moderate-resolution Imaging Spectroradiometer (MODIS) NDVI data (MOD13Q1) was used for medium scale vegetation trend analysis. Harmonic analyses and non-parametric trend tests were applied to both GIMMS NDVI (1981–2006) and MODIS NDVI (2001-2011) data sets. Based on a robust trend

[7] *This chapter is based on* Teferi, E., Uhlenbrook, S., Bewket, W. (2014). Inter-annual and seasonal trends of vegetation conditions in the Upper Blue Nile (Abay) basin: Dual scale time series Analysis. **Earth Syst. Dynam. Discuss.**, 6, 267-315. doi:10.5194/esdd-6-169-2015.

estimator (Theil-Sen slope) most part of the UBN (~77%) showed a positive trend in monthly GIMMS NDVI with a mean rate of 0.0015 NDVI units (3.77% yr^{-1}), out of which 41.15% of the basin depicted significant increases ($p < 0.05$) with a mean rate of 0.0023 NDVI units (5.59% yr^{-1}) during the period. However, the MODIS-based vegetation trend analysis revealed that about 36% of the UBN shows a significantly decreasing trend ($p < 0.05$) over the period 2001-2011 at an average rate of 0.0768 NDVI yr^{-1}. This indicates that the greening trend of vegetation condition was followed by browning trend since the mid-2000s in the basin, which requires the attention of land users and decision makers. Seasonal trend analysis was found to be very useful in identifying changes in vegetation condition that could be masked if only inter-annual vegetation trend analysis was performed. Over half (60%) of the Abay basin were found to exhibit significant trends in seasonality over the 25 years period (1982-2006). About 17% and 16% of the significant trends consisted of areas experiencing a uniform increase in NDVI throughout the year and extended growing season, respectively. These areas were found primarily in shrubland and woodland regions. The MODIS-based trend analysis revealed trends that were more linked to human activities. This study concludes that integrated analysis of inter-annual and intra-annual trends based on GIMMS and MODIS enables a more robust identification of changes in vegetation condition.

9.2   Introduction

Land degradation is a widespread environmental and development challenge (e.g.Dregne et al., 1991; UNEP, 2007). It is central to many international conventions and protocols related to environmental protection. Increasing demands for food, water and energy resulting from the growth in population and *per capita* consumption are driving unprecedented land use change (Godfray et al., 2010; Kearney, 2010). In turn, unsustainable land use is causing degradation of land resources. Thus, up-to-date quantitative information about land degradation is crucial to develop sustainable land use strategies and to support policy development for food and water security and environmental integrity. Status and trend of vegetation condition generally serve as a proxy for land degradation (Metternicht et al., 2010; Wessels et al., 2004; Wessels et al., 2007).

Detecting and characterizing trends in vegetation condition over time using remotely sensed data has received considerable attention in recent years (Bai et al., 2008; de Jong et al., 2011; Eastman et al., 2013; Verbesselt et al., 2010a). Recent interest in vegetation trend analysis arises for three reasons. First, there is considerable interest in monitoring and assessing the state and trend of land degradation as well as for monitoring the performance of management programs (Buenemann et al., 2011; Vogt et al., 2011). Second, it is only recently that substantial amounts of remotely sensed data and robust geospatial approaches suitable to such analysis are becoming available (Bai et al., 2008; Buenemann et al., 2011; Tucker et al., 2005). Third, it is a natural first step toward identifying drivers of changes in terrestrial ecosystems as vegetation variability and trends affect the exchange of water, energy, nutrients and carbon between the biosphere, the geosphere and the atmosphere (Baldocchi et al., 2001).

Changes in vegetation occur in three ways: (1) a seasonal or cyclic change that is driven by climate (e.g. annual temperature and rainfall) impacting plant phenology; (2) a gradual change over time that is consistent in direction (monotonic) such as change in land management or land degradation; and (3) an abrupt shift at a specific point in time (step trend) that may be caused by disturbances such as a sudden change in land use policies, deforestation, floods, droughts, and fires (Angert et al., 2005; de Jong et al., 2013a; Slayback et al., 2003; Tucker et al., 2001; Verbesselt et al., 2010b). Thus, considering all types of changes (i.e. seasonal, gradual and abrupt changes) is essential in order to assess the environmental impact of vegetation changes or to be able to attribute the changes in vegetation to drivers behind (de Jong et al., 2013a; Verbesselt et al., 2010b). There is substantial interest in monitoring the trends in seasonality of vegetation due to its sensitive response to climate change (Parmesan and Yohe, 2003; Sparks et al., 2009; Walther et al., 2002).

Long-term, remotely sensed normalized difference vegetation index (NDVI) data are suitable to detect and characterize trends in vegetation condition over time (de Jong et al., 2011; Eastman et al., 2013; Tucker et al., 2001; Verbesselt et al., 2010a). Although NDVI can be computed from different multispectral satellite data, NDVI from Advanced Very High Resolution Radiometer (AVHRR) is the only global vegetation dataset which spans a time period of three decades. Hence, it allows quantification of ecosystem changes as a result of ecosystem dynamics and varying climate conditions. In this study, AVHRR based Global Inventory, Monitoring, and Modelling Studies (GIMMS) NDVI data sets (1981–2006) were used for the purpose of long-term trend analysis (Tucker et al., 2005). Besides, NDVI time series (2001-2011) from the Moderate resolution Imaging Spectroradiometer (MODIS) 250m (MOD13Q1) on board the Earth Observing System Terra platform were used for medium scale vegetation trend analysis.

Several vegetation trend studies from analysis of satellite observations have reported increasing trend of greenness in the Northern Hemisphere, including the Sahel (Fensholt et al., 2009; Karlsen et al., 2007; Slayback et al., 2003; Tucker et al., 2001). Other studies based on phenology and temperature observations from stations also confirmed greenness trends (Sparks et al., 2009). However, Zhao and Running (2010) reported a decreasing trend in greenness globally during the period 2000-2009 using Moderate-resolution Imaging Spectroradiometer (MODIS) NDVI data. Similarly, de Jong et al. (2011) found that many forested biomes experienced a decline in vegetation greenness. There is still no consistent result on trends of global vegetation activity (Bala et al., 2013).

The inconsistent results from various studies potentially emanated from differences in location of study areas (variations in altitude and latitude), trend detection techniques (ordinary least square versus median trend), the length of the data series and NDVI data sources used (e.g. GIMMS, MODIS). Therefore, it is very important to characterize and understand inter-annual and intra-annual variability of vegetation activity using long-term data sets in study areas such as Abbay basin (1982-2006) where climate variability is high and topography is diverse.

It is essential to characterize and understand inter-annual and intra-annual variability of vegetation activity using long-term data sets such as GIMMS NDVI, especially, if the increase or decrease of vegetation growth in Abay basin is mainly driven by climatic factors (De Beurs et al., 2009). However, if vegetation growth is mainly caused by human activities such as sustainable land management (SLM) programs that involve localized implementation of best management practices (BMPs), then it is useful to use MODIS NDVI data as it provides repeated information at the spatial scale at which most human-driven land cover changes occur (De Beurs et al., 2009; Gallo et al., 2005; Townshend and Justice, 1988). Such analysis reveals areas of change that merit closer attention since it depicts significant shifts in local water, carbon and energy fluxes (Henebry, 2009; Vuichard et al., 2008). The results of this study are very important from context of Upper Blue Nile (UBN) basin for two reasons. First, the UBN basin is the major water source area of the Nile river, as it contributes around 70% of the overall Nile flow. Thus, the UBN basin is the most important river basin not only for Ethiopia, but also for the downstream Nile basin countries (i.e. Sudan and Egypt). Second, the UBN basin accounts for a major share of the country's irrigation and hydropower potential.

The general objective of this study was to characterize the degree of improvement or degradation of vegetation condition in the Abay basin using both coarse and medium scale analysis. The specific objectives were: (1) to identify inter-annual and seasonal trends in vegetation conditions at both coarse and medium scales using robust trend estimators; and (2) to detect trend breaks.

9.3 MATERIALS AND METHODS

9.3.1 Study area and data Sets

The Upper Blue Nile/Abbay basin covers a total area of 199,812 km^2 and is located in the center and west of Ethiopia (Fig. 1). It lies approximately between $7°45'$ and $12°46'$ N, and $34°06'$ and $40°00'$ E. Altitude of the Upper Blue Nile basin ranges from 475 m asl at the Sudanese border to 4,257 m asl at the summit of Mt. Guna. More than 83% of the basin is located at an elevation above 1,000 m asl. The multi-year (1983-2006) mean annual rainfall varies between about 894 mm a-1 to 1,909 mm a-1, with a spatial mean of about 1,396 mm a-1 based on gridded station-based rainfall data. The multi-year (1982-2006) mean annual areal potential evaporation ranges from 1065 mm a-1 to 1756 mm a-1, with a spatial mean of about 1,396 mm a-1 based on the Penman–Monteith method from Climate Research Unit 3.10 data set (Harris et al., 2014). The multi-year (1982-2006) annual average temperature varies across space from 14°C to 29°C based on CRU 3.10 data set.

Three types of data sets were used in this study: (1) GIMMS 15-day composite NDVI product, (2) MODIS NDVI 16-Day L3 Global 250m (MOD13Q1) and (3) Land use and land cover (LULC) data produced by Woody Biomass Inventory and Strategic Planning Project (WBISPP) at a scale of 1:250,000 (TECSULT, 2004). The Global Inventory, Monitoring, and Modelling Studies (GIMMS) NDVI data product (Tucker et al., 2005), was used to analyse the long-term (1982-2006) vegetation variability and trends in UBN basin. The MODIS 16-day composite NDVI data sets with 250 m spatial resolution (MOD13Q1) tiles of h21v07 and h21v08 for the period 2001–2011 obtained from the NASA Earth Observing System (EOS) data gateway were used in this study (Huete et al., 2002). The MODIS NDVI complements NOAA's AVHRR derived NDVI products and provides continuity for time series historical applications.

9.3.2 Harmonic Analysis of NDVI Time-series

Both AVHRR and MODIS NDVIs are composite products. Since Maximum Value Composite (MVC) is not an atmospheric-correction method, some artefacts related to residual cloud cover and atmospheric haze remain in the MVC processed NDVI series (Nagol et al., 2009). Therefore, it was essential to perform harmonic analysis of both GIMMSg and MDOIS NDVI data sets using HANTS (Harmonic ANalysis of Time Series) algorithm (Roerink et al., 2000; Verhoef et al., 1996) for the purpose of: (i) screening and removal of cloud contaminated observations; and (ii) temporal interpolation of the remaining observations to reconstruct gapless images at a prescribed time. Harmonic (Fourier) analysis has been successfully applied to describing precipitation patterns (e.g.Landin and Bosart, 1989), meteorology (e.g.Legates and Willmott, 1990) and seasonal and interannual variations in land surface condition (Jakubauskas et al., 2001). The HANTS algorithm was developed based on the concept of Discrete Fourier Transform. HANTS considers only the most significant frequencies expected to be present in the time profiles, and applies a least squares curve fitting procedure based on harmonic components (sine and cosine). Thus, harmonic (Fourier) analysis decomposes a time-dependent periodic phenomenon (e.g. vegetation development) into a series of sinusoidal functions, each defined by unique amplitude (strength) and phase (orientation with respect to time) (Roerink et al., 2000;Verhoef et al., 1996) and can be described as the sum of sine and cosine components as follows (Harris and Stöcker, 1998; Wilks, 2011):

$$f(t) = \alpha_0 + \sum_{n=1}^{n=2} C_n \cos(\omega_n t - \varphi_n) \tag{9.1.a}$$

$$= \alpha_0 + \sum_{n=1}^{n=2} \left\{ a_n \sin\left(\frac{2\pi n t}{T}\right) + b_n \cos\left(\frac{2\pi n t}{T}\right) \right\} + e \tag{9.1.b}$$

where $f(t)$ is the series value at time t, T is the length of the series (e.g. T=12 observations for a monthly time series), n is the number of harmonics to be used in the regression (in this study 2 harmonics were used), α_0 is the mean of the series, C_n is the amplitude of harmonics given by $C_n = \sqrt{a_n^2 + b_n^2}$, ω_n is the frequency $(2\pi n/T)$, φ_n is the phase angle given by $\varphi_n = \tan^{-1}(b_n/a_n)$ and e is the error term. Eq. (9.1.a) emphasizes the harmonic interpretation, while Eq. (9.1.b) is a convenient transformation to a multiple linear harmonic regression model with coefficients $a_n = C_n \cos(\varphi_n)$ and $b_n = C_n \sin(\varphi_n)$ that can be easily estimated.

In this study the IDL implementation of the HANTS algorithm (De Wit and Su, 2005) was used. Essential HANTS parameters were set as in Table 43 in order to obtain a reliable fitting curve. Detail explanation about the parameters can be found in Roerink et al. (2000). The zero frequency (mean annual NDVI) and the frequencies with time periods of 1 year (annual) and 6 months (semi-annual) were selected for the analysis. So, the output comprises five Fourier components (i.e.3 amplitudes and 2 phases).

Table 43 Parameters used in HANTS analysis

HANTS parameters	GIMMS		MODIS	
	Single year	Full data set	Single year	Full data set
Number of frequency	0, 1, 2	0, 24, 48	0, 1, 2	0, 23, 46
Invalid data rejection threshold:	0 - 1	0 - 1	0 - 1	0 - 1
Fit error tolerance (FET)	0.1	0.1	0.1	0.1
Maximum iterations (i_{MAX})	6	12	6	12
Minimum retained data points	16	416	15	165

9.3.3   Long-term Trend Analysis

Before the time series analysis the biweekly HANTS filtered GIMMS and MODIS data were aggregated to monthly composites by applying the arithmetic mean of each month. Seasonal variation must be removed in order to better discern the long-term trend in NDVI over time. Thus, a seasonally adjusted series is needed in order to better discern any trend that might otherwise be masked by seasonal variation. Standardized anomalies (Z-score) of the monthly HANTS filtered NDVI data were calculated to remove the seasonality from the original time series (i.e. *de-seasoning*) using the following equation:

$$Z = (x - \mu)/\sigma \tag{9.2}$$

where x is the data value of the respective month, μ is the monthly long term average value (i.e., the long term average of all Januaries, Februaries, ...) and σ is the monthly standard deviation. A Z-score value of 0 would mean that it has a value equal to the long-term mean; a value of +1 would mean that it is 1 standard deviation of the long term mean, and so on.

The de-seasoned monthly NDVI time-series were corrected for serial correlation (autocorrelation) using a Trend Preserving Prewhitening (TPP) approach (Wang and Swail, 2001). The prewhitened series have the same trend as the original series, but with no serial correlation. Three types of inter-annual trend analysis and one trend significance test were applied to pre-whitened data on a pixel basis: linear trend (ordinary least squares regression (OLS)), median trend (Theil-Sen), monotonic trend (Mann-Kendall) and Mann-Kendall significance. The presence of a linear trend was tested by both linear trend (OLS) and median trend (Theil-Sen) operators. The linear trend (OLS) trend operator computes the slope coefficient of an ordinary least squares regression between the values of each pixel over time and a perfectly linear series. If a plot of NDVI data (Y) versus time (t) suggests a simple linear increase or decrease over time, a simple linear regression of Y on time t is a test for linear trend: $Y = \beta_0 + \beta_1 t + e$ where β_0 is the intercept, β_1 is the slope coefficient, and e is the error term. The t-statistic may be used to determine if the slope coefficient $\beta 1$ is significantly different from zero. However, the slope computed in this way is sensitive to outliers in the data and the data need to be normally distributed. Therefore, a robust non-parametric operator, the median trend (Theil-Sen (TS)) was computed for assessing the linear trend and rate of change per unit time (Hoaglin et al., 1983). Theil-Sen (TS) slope estimator was proposed by Thiel (1950) and modified by Sen (1968) and computes the median of the slopes between observation values at all pairwise time steps given by the following equation:

$$\beta_1 = median\left(\frac{y_j - y_i}{x_j - x_i}\right) \tag{9.3}$$

for all $i < j$ and $i = 1, 2, ..., n\text{-}1$ and $j = 2, 3,..., n$

The major advantage of TS slope estimator is because of its breakdown bound. The breakdown bound for a robust statistic is the number of wild values that can occur within a series before it will be affected. For the median trend, the breakdown bound is approximately 29%, meaning that the trends expressed in the image must have persisted for more than 29% of the length of the series (in time steps). For example, in this study the time series data contains 300 monthly images so it is completely unaffected by wild and noisy values unless they persist for more than 87 months (i.e.0.29*300=7.25 years). The implication of this is that it also ignores the effects of short-term inter-annual climate teleconnections such as El Niño (typically a 12 months effect) and La Niña (typically a 12-24 months effect).

The non-linear trend was tested by mapping the degree of trend monotonicity (using the Mann-Kendall statistic) − the degree to which the trend is consistently increasing or decreasing. The significance of the trend was assessed using the non-parametric Mann-Kendall test.

9.3.4 Seasonal Trend Analysis

Seasonal trend analysis was performed in order to identify trends in the essential character of the seasonal cycle while rejecting noise and short-term variability seasonal cycle (Eastman et al., 2013). This was performed based on a two-stage time series analysis. First, harmonic regression is applied to each year of images in the time series to extract an annual sequence of overall greenness and the amplitude and phase of annual and semi-annual cycles (Eastman et al., 2009). This suite of harmonics is termed as greenness parameters. Amplitude 0 represents the annual mean NDVI or overall greenness for each year. Amplitude 1 represents the peak of annual greenness. Phase 1 denotes the timing of annual peak greenness, represented by the position of the starting point of the representative sinewave of annual greenness. An increase or decrease in the phase angle means a shift in the timing to an earlier or later time of the year, respectively. The values of Phase image potentially ranges from 0 degree to 359 degrees such that each 30 degrees indicates a shift of approximately one calendar month (Eastman et al., 2009). Second, trends over years in the greenness parameters were analyzed using a Theil-Sen median slope operator. This procedure is robust to short-term interannual variability up to a period of 29% of the length of the series (Eastman et al., 2013; Gilbert, 1987; Hoaglin et al., 1983). Furthermore, the samples are annual measures of shape parameters. Thus, short inter-annual disturbances (*i.e.*, those 8 years or less in this case) such as individual El Niño/Southern Oscillation events, have little effect on the long-term trends represented by the median trends. The significance of the trends in the five shape parameters was tested using the Contextual Mann Kendall statistics (CMK) (Neeti and Eastman, 2011) which is a modified form of the Mann-Kendall test of trend significance (Kendall, 1948; Mann, 1945). CMK trend test reduces the detection of spurious trends and an amplification of confidence when consistent trends are present through adding contextual information. CMK is a modified version of the Mann-Kendall test which is based on a principle that a pixel would not be expected to exhibit a radically different trend from neighboring pixels.

Amplitude trend image was created by color compositing trends in Amplitude 0 in red, Amplitude 1 in green and Amplitude 2 in blue. Similarly, phase trend image was created by compositing Amplitude 0 in red, Phase 1 in green and Phase 2 in blue. The colors in the amplitude trend and phase trend images can be used as a guide to locating areas that are going through similar trends in seasonality. Furthermore, in order to visualize the changes in seasonality, fitted curves for the beginning and the end of the series were developed based on the trends determined from the whole series using a Theil-Sen median slope operator.

In order to make the interpretation easier, the seasonal trends were categorized into different classes based on major CMK trend significance. The CMK trend significance for each of the five shape parameters was classified into three major categories: significantly decreasing at the $p < 0.05$ level, not significant, and significantly increasing at the $p < 0.05$ level. All combinations of these significance categories over the trends of five shape parameter (Amplitude 0, Amplitude 1, Amplitude 2, Phase 1 and Phase 2) produced a total of 54 seasonal trend classes. Each class is defined by a combination of five independent tests. Only five major seasonal trend classes in terms of area were selected for detailed analysis.

9.3.5   Change Point Analysis

A change-point analysis was conducted, because it indicates that the trends change between positive and negative within the analysis period. Regions of interest (ROIs) were selected based on most prevalent classes of significant seasonal trends for GIMMS NDVI and MODIS NDVI. The Pettitt test of change-point (Pettitt, 1979) and Breaks For Additive Seasonal and Trend (BFAST) method (Verbesselt et al., 2010a; Verbesselt et al., 2010b) were used to detect and characterize abrupt changes within the trend and seasonal components of the ROI times series. The Pettitt test is a nonparametric statistics used to identify a single change point and less sensitive to outliers (Wijngaard et al., 2003). It is extensively employed in change detection studies (Gao et al., 2011; Guerreiro et al., 2014; Villarini et al., 2011). However, if long-term trends are composed of consecutive segments with gradual change, separated from each other by a relatively short period of abrupt change, Pettitt test cannot handle it (de Jong et al., 2012; Verbesselt et al., 2010b). Thus, the use of BFAST trend break analysis assisted identification of several change periods which might otherwise been overlooked through a commonly assumed fixed change trajectory analysis. BFAST is an iterative algorithm that integrates the decomposition of time series into seasonal, trend, and remainder component with methods for detecting changes. An additive decomposition model is used to iteratively fit a piecewise linear trend and a seasonal model. The general model is of the form: $Y_t = T_t + S_t + e_t,\ (t \in m)$, where Y_t is the observed NDVI data at time t (in the time series m), T_t is the trend component, S_t is the seasonal component, and e_t is the remainder component. A trend component corresponds to a long-term process that operates over the time spanned by the series. A remainder component corresponds to a local process which causes variability between cycles. A seasonal component corresponds to a cyclic process that operates within each cycle.

9.4 RESULTS

9.4.1 Inter-annual trends in NDVI

GIMMS-based inter-annual trends

The Upper Blue Nile (Abay) basin is characterized by a multi-year mean NDVI of 0.49 for the period 1982–2006 using the GIMMS data set (Fig. 42(a)). On the one hand, the south and southwest region of the basin represents the highest mean NDVI (Fig. 42(a)) and the lowest temporal variation variation (Fig. 42(b)). This area is characterized by shrubland and woodland vegetation. On the other hand, the eastern, northeastern and northwestern part represents the lowest mean NDVI and the highest temporal variation as a consequence of the relatively low annual precipitation and the more extended dry periods. Besides, this area is characterized by intensive and sedentary cultivation.

During the period 1982-2006, based on median trend about 77% of the basin showed a positive trend in monthly NDVI with a mean rate of 0.0015 NDVI units (3.77% yr^{-1}) and the remaining 23% of the basin exhibited a decreasing trend with a mean rate of 0.0007 NDVI units (1.91% yr^{-1}) (Fig. 42(a) and Table 44). The monthly NDVI increased significantly (p<0.05) over 41.15% of the basin area, with a mean rate of 0.0023 NDVI units (5.59% yr^{-1}) and decreased significantly over 4.6% of the basin, with a mean rate of 0.0017 NDVI units (4.44% yr^{-1}).

Table 44 Result of linear trend analysis based on Theil-Sen (TS) (median) slope for the monthly GIMMS NDVI data

Trend estimator	Slope		MK-Z trends	
	Positive	Negative	Significantly +	Significantly -
ΔNDVI/month	0.0015	0.0007	0.0023	0.0017
% NDVI change/month	0.314	0.159	0.466	0.370
Land area (%)	76.85	23.15	41.15	4.60

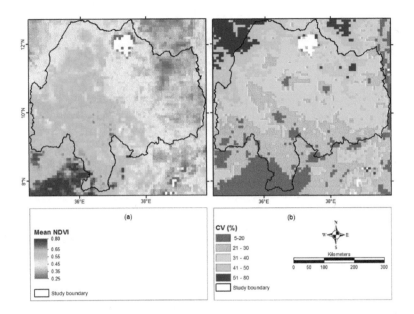

Figure 42 Spatial distribution of multi-year (1982-2006) mean monthly GIMMS NDVI (a) and coefficient of variation (b)

Figure 43 depicts the impact of assuming a linear trend development in the 25 years NDVI data series. Figure 43(c) clearly shows the spatial distribution of the difference between the linear trend and non-linear trend for areas of positive NDVI trend values with mean difference value of 0.043 indicating that the greater part of the basin is described better using a linear trend measure as compared to non-linear trend measure. However, the MK test values in the southern and eastern part of the basin attained higher values as compared to the linear test. For areas with negative NDVI changes, it appears that the MK test described the NDVI trends better as compared to the LM test with a mean difference value of -0.03 (Fig. 43(d)).

158

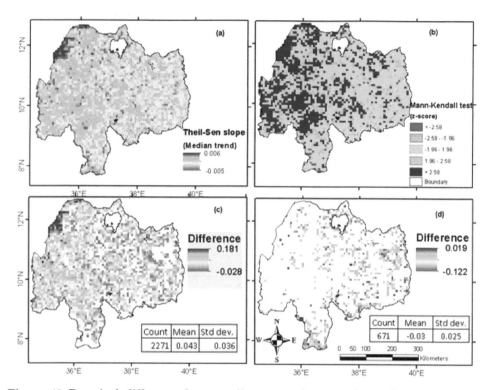

Figure 43 Per-pixel difference between linear trend test and non-linear monotonic (Mann–Kendall) trend test on NDVI in Abay (UBN) basin for the period 1982 to 2006): (a) median trend (b) Mann-Kendall significance test z-score, (c) the spatial distribution map of per-pixel difference between linear trend test and non-linear monotonic (Mann–Kendall) trend test for areas of positive NDVI trends, and (d) the spatial distribution map of per-pixel difference between linear trend test and non-linear monotonic (Mann–Kendall (MK)) trend test for areas of negative NDVI trends. MK test values were subtracted from the linear test values

MODIS-based inter-annual trends

Areas of higher multi-year mean NDVI depict lower vegetation cover variability (Fig.44(a) and Fig. 44(b)). Areas located in the eastern part of the basin are those with higher variability and show an upward (positive) trend of vegetation growth in the period 2001-2011 (Fig.44(b) and Fig.44(c)). About 36% of the UBN/Abay basin shows a significantly browning trend ($p < 0.05$) over the period 2001-2011 at an average rate of 0.0768 NDVI yr^{-1} using MODIS 250 m spatial resolution (Table 45 and Table 46). By contrast, only about 1.19% of the basin shows a significantly greening trend ($p < 0.05$) over the period 2001-2011 at an average rate of 0.066 NDVI yr^{-1} (Table 45 and Table 46). The majority of this browning trend comes from Dabus (14.27% at a rate of 0.0744), Belles (13.5% at a rate of 0.0816), Wonbera (11.81% at a rate of 0.0792) and Didessa (11.48% at a rate of 0.0792). The vegetation activity in Belles sub basin is being decreased at a faster rate (0.0816 NDVI yr^{-1}) than the others (Table 45 and Table 46).

Table 455 The percentage of sub-basin's significant trend out of the total area of significant trend in the basin and the percentage of area significantly increasing or decreasing within sub-basins are shown. The locations of the sub-basins can be found in Fig. 45

Sub-basins	% of total trend			% of sub-basin's area		
	Insignificant trend	Significantly greening	Significantly browning	Insignificant trend	Significantly greening	Significantly browning
Anger	3.37	0.46	5.27	52.96	0.14	46.90
Belles	3.77	1.12	13.50	32.90	0.18	66.91
Beshilo	9.78	18.23	0.97	91.65	3.22	5.13
Dabus	8.76	5.07	14.27	51.68	0.56	47.76
Didessa	9.28	1.76	11.48	58.64	0.21	41.15
Dinder	6.78	0.18	9.19	56.54	0.03	43.43
Finca	2.66	2.96	0.96	81.62	1.71	16.67
Guder	4.85	7.61	1.15	85.87	2.54	11.59
Jemma	11.48	22.03	1.45	90.25	3.26	6.49
Muger	5.88	12.15	0.86	89.13	3.47	7.40
North Gojam	8.65	12.78	4.76	74.64	2.08	23.28
Rahad	2.64	0.01	7.10	39.59	0.00	60.41
South Gojam	8.43	3.88	8.84	62.37	0.54	37.09
Tana	5.11	1.89	8.11	52.45	0.37	47.18
Welaka	4.82	9.77	0.29	93.24	3.56	3.20
Wonbera	3.74	0.10	11.81	35.86	0.02	64.12
Total	100	100	100	63.05	1.19	35.76

The Maximum significant browning slope observed in Didessa sub-basin is the largest of all maximum slopes in the basin. The size of greening and the browning area within sub-basins are proportional in sub-basins such as Beshillo, Jemma, Muger and Welaka. Whereas, more than half of the areas of the Belles (66.91%), Rahad (60.41%), and Wonbera (64.12%) are represented by significant browning trend. The majority of this greening trend comes from Jemma (22.03% at a rate of 0.066), Beshillo (18.23% at a rate of 0.066), North Gojam (22.78% at a rate of 0.066), and Muger (12.15% at a rate of 0.0648) sub-basins. The slowest rate and the fastest rate of significant greening are observed in Dinder sub-basin (0.0552 NDVI yr^{-1}) and Dabus sub-basin (0.0768 NDVI yr^{-1}), respectively.

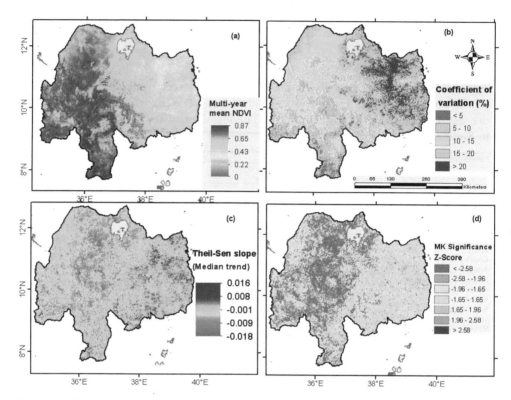

Figure 444 Vegetation cover variability and trends for the Upper Blue Nile (Abay) basin based on MODIS 250m NDVI (MOD13Q1) for the period (2001-2011): (a) multi-year mean monthly NDVI; (b) coefficient of variation of monthly NDVI; (c) Theil-Sen slope (ΔNDVI/month); and (d) Mann-Kendall (Shukla et al.) significance test of the trend test

Table 466 Summary statistics of Theil-Sen slope (median trend) for the significantly increasing (greening) and decreasing (browning) trends per sub-basins. Locations of the sub-basins can be found in Fig. 45

Sub-basins	Significantly greening (Δ NDVI yr^{-1})			Significantly browning (Δ NDVI yr^{-1})		
	Minimum	Maximum	Mean	Minimum	Maximum	Mean
Anger	0.0444	0.0984	0.0612	0.0336	0.1740	0.0768
Belles	0.0456	0.1524	0.0744	0.0348	0.1908	0.0816
Beshilo	0.0420	0.1596	0.0660	0.0384	0.1572	0.0660
Dabus	0.0432	0.1812	0.0768	0.0372	0.1764	0.0744
Didessa	0.0444	0.0984	0.0624	0.0312	0.2124	0.0792
Dinder	0.0372	0.0912	0.0552	0.0348	0.1692	0.0756
Fincha	0.0444	0.1524	0.0732	0.0372	0.1656	0.0708
Guder	0.0420	0.1176	0.0648	0.0360	0.1584	0.0684
Jemma	0.0408	0.1920	0.0660	0.0348	0.2016	0.0660
Muger	0.0432	0.1224	0.0648	0.0408	0.1464	0.0684
North Gojam	0.0372	0.1332	0.0660	0.0384	0.1704	0.0732
Rahad	0.0528	0.0660	0.0600	0.0384	0.1620	0.0756
South Gojam	0.0408	0.1308	0.0636	0.0356	0.1836	0.0768
Tana	0.0444	0.1476	0.0672	0.0372	0.2028	0.0792
Welaka	0.0348	0.1476	0.0660	0.0420	0.1488	0.0636
Wonbera	0.0504	0.1452	0.0696	0.0336	0.1884	0.0792
Total	0.0348	0.1920	0.0660	0.0312	0.2124	0.0768

9.4.2   Seasonal trends

GIMMS-based seasonal trends

More than half (59.52%) of land areas exhibit a significant trend in seasonality over the 25 year time period measured (1982-2006), producing a total of 53 significantly trending classes of seasonal curves which include different combinations of significant changes in Amplitude 0, Amplitude 1, Amplitude 2, Phase 1 and Phase 2. Thus, a trend change in seasonality is not a rare occurrence in Abay basin. Out of the 53 classes, only the prominent five classes of seasonal trends are described below, which account for about 67% of all significantly trending pixels and represent 39.95% of all land pixels of the basin.

Class 1 has an area of 19,712 km^2 which accounts for 16.85% of all significantly trending areas and 10.03% of the total land area (Table 47 and Fig. 45). This class represents areas with a significantly positive trend in Amplitude 0 (mean annual NDVI) and no significant change in any of the other four parameters (Amplitudes, Phases 1 and Phase 2), producing curves that are raised uniformly throughout the year. The frequency distribution of seasonal trend classes within zones of land cover types shows that the majority of Class 1 trending pixels occur in areas of woodland

162

(33%) and shrubland (30%). Thus, it appears that woodland and shrubland areas are experiencing a consistent increase in productivity throughout the year.

Table 477 The most important land cover classes for each major significant seasonal trend categories

Classes of seasonal trends	Grassland	Shrubland	Woodland	Bare land	Crop land	Natural Forest	Total
Class 1 (Sig. ↑ in A0)	3072 (16)	5888 (30)	6528 (33)	448 (2)	3776 (19)	0 (0)	19712 (100)
Class 2 (Sig. ↑ in A0 & ↓ in P1)	704 (4)	3712 (20)	10304 (56)	256 (1)	3584 (19)	0 (0)	18560 (100)
Class 3 (Sig. ↑ in A1)	960 (14)	704 (10)	448 (7)	64 (1)	4672 (68)	0 (0)	6848 (100)
Class 4 (Sig. ↓ in A1)	3520 (19)	1792 (10)	1024 (5)	1216 (7)	11008 (59)	128 (1)	18688 (100)
Class 5 (Sig. ↓ in P1)	576 (4)	2368 (16)	7040 (48)	128 (1)	4352 (30)	256 (2)	14720 (100)
Not significant	14144 (18)	11328 (14)	13568 (17)	4160 (5)	35008 (44)	1344 (2)	79552 (100)
Others (Small sig. changes)	5376 (14)	5504 (14)	7872 (20)	2112 (5)	16576 (43)	1024 (3)	38464 (100)

Class 2 is characterized by both a significant increase in Amplitude 0 (mean annual NDVI) and a significant decrease in Phase 1 (orientation with time of the annual cycle). A decreasing phase angle means a shift to a later time of the year. This class has an area of 18,560 km^2 and describes about 10% of the basin land area (Table 47 and Fig. 45). Figure 46(b) shows that the green-down period is coming about 15 days later in 2006 than 1982. About 56% of Class 1 trending pixels occur in areas of woodland. This means that woodland areas of the Upper Blue Nile/Abay basin are experiencing an increase in mean annual NDVI with significant shift in end of growing season and no effect on the start of growing season leading to lengthening of the growing season.

Class 3 represents areas that exhibit a significant increase in Amplitude 1 (the annual cycle), but no other significant trend parameters. Class 3 covers 3.48% (6,848 km^2) of all the basin area (Table 47 and Fig. 45). This class of seasonal trend exhibits a characteristic of increase in NDVI during the growing season (Aug-Nov) balanced by declines during the dry season (Feb-May) in order to maintain the same mean NDVI (Fig.46(c)). The majority of Class 3 trending pixels occur in cropland areas (68% or 4,672 km^2) of the basin. Therefore, NDVI is enhanced during the green season (cropping season) and inhibited during the brown season or harvest.

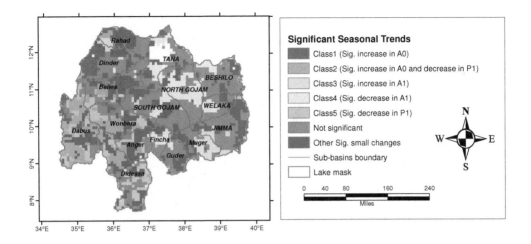

Figure 455 The five most prevalent classes of significant seasonal trends in GIMMS NDVI (1982-2006): Class 1 shows significant increases in Amplitude 0 (Mean NDVI); Class 2 shows significant increases in Amplitude 0 and decrease in Phase 1 (the timing of annual peak greenness) together; Class 3 shows significant increases in Amplitude 1 (increase in the difference between minimum and maximum NDVI without affecting the mean); Class 4 shows significant decreases in Amplitude 1; and Class 5 shows significant decreases in Phase 1.

Class 4 represents areas that exhibit a significant decrease in Amplitude 1 (the annual cycle), but no other significant trend parameters. Class 4 covers 9.51% (18,688 km^2) of all the basin area (Table 47). This class of seasonal trend exhibits a characteristic decrease in NDVI during the growing season (more pronounced between Aug-Nov) increase in NDVI during dry season (Fig. 46(d)). Class 4 seasonal trend pixels are mostly associated with cropland (59%) and grassland (19%) land cover types. This type of seasonal trend is more common in North Gojam and Lake Tana sub-basins.

Class 5 describes areas that show a significant decrease in Phase 1 only. Class 5 seasonal curves resemble Class 2 seasonal curves in that both have similar pattern of a decrease in yearly phase. It covers 7.5% (14 720 km^2) of all the basin area and represents 12.6% of all significant seasonal trend pixels. This class of seasonal trend exhibits a characteristic shift in the end of growing season. Class 5 pixels are mostly associated with woodland (48%) and cropland (30%). This type of seasonal trend is more common in southern and southwestern part of the basin.

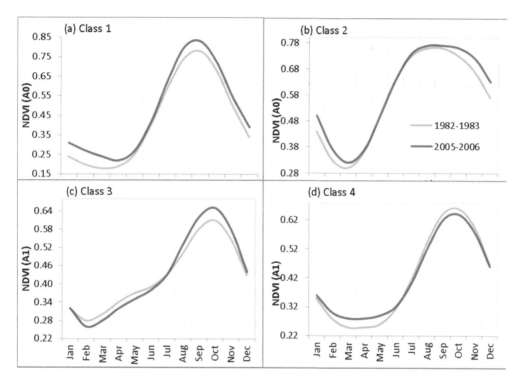

Figure 466 Seasonal curves representative of each of the major trend classes from the 1982–2006 series: the green curve represents the characteristic seasonal curve at the beginning of the series (mean value of 1982 and 1983) while the red curve represents the characteristic seasonal curve at the end of the series (mean value of 2005 and 2006).

MODIS-based seasonal trends

About 207 classes of significant intra-annual changes (prominent classes are depicted in Figure 47 (a)), that include different combinations of significant changes in Amplitude 0, Amplitude 1, Amplitude 2, Phase 1 and Phase 2, were found. The MODIS-based analysis revealed that about 96% of the basin represents significant seasonal trends, out of which 33.15% represents the dominant classes. About 46% of the dominant classes represent areas with vegetation that underwent significant decrease in the annual vegetation cycle (A1) or both the annual vegetation cycle (A1) and mean annual vegetation cycle (A0) (Fig. 47). These areas are represented by ROI#2, ROI#5 and ROI#8 (Fig. 47). Fig. 48(b) and 48(e) show a change in the greenness pattern from a strong mean annual and annual cycle of these areas to strong semi-annual vegetation cycle. This change was found to be statistically significant ($p < 0.05$) based on the CMK test. This means these areas underwent significant changes from shrubland/woodland vegetation to double cropping (irrigation) during the period 2001-2011. For example, ROI#2 and ROI#5 located in the light green color of Fig. 47 represents small-scale irrigation areas with strong semi-annual cycle in Didessa sub-basins. ROI#8 indicates that the vegetation in this area has changed from strong annual mean and annual NDVI to a strong semi-annual vegetation cycle. This is a typical characteristic of change in land use from shrubs and bushes to irrigated cropland.

The Magenta color is dominant in Muger (ROI#1), Guder and North Gojam sub-basins and it represents a significant change from strong annual vegetation cycle to strong mean annual and semi-annual vegetation cycles. The western part of the basin is dominated by yellow color (ROI#6), which means that the cycle of vegetation has undergone a change from strong semi-annual to strong mean annual. Cyan color is wide spread, particularly in Didessa sub-basin (ROI#3), meaning that currently the vegetation in this area is characterized by a significant decrease in mean annual vegetation cycle indicating degradation of forest without complete removal. Figure 48 (c) shows a decrease in vegetation condition in the sampled area, while the NDVI still remain above 0.8.

The UBN/Abay basin is not only characterized by a significant decreasing seasonal trend of vegetation but also increasing trend. The increasing seasonal trends are mostly found, in the eastern part of the basin, where previous de-vegetation took place. For example, the Jema (ROI#7) sub-basin is one of the areas that a significantly increasing mean annual NDVI can be observed. In other words, there has been a change from a strong semi-annual vegetation cycle to strong mean annual NDVI. This could be related to increased vegetation activity in the area (e.g. on-farm eucalyptus plantation). ROI#4 is also another example for a significant increasing trend. ROI#4 represents a significant change from strong annual vegetation cycle to strong mean annual and semi-annual vegetation cycles. This is due to small-scale irrigation development on the area mainly because of the change in river course of Megech, North of Lake Tana.

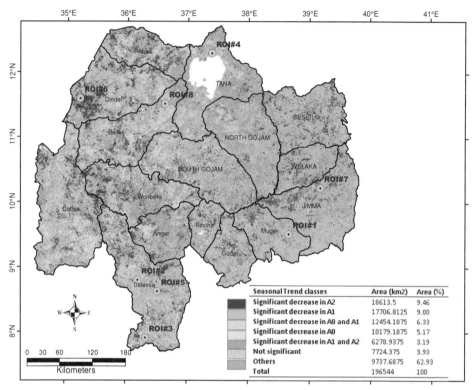

Figure 477 Most prevalent classes of significant seasonal trends of monthly 250m MODIS NDVI (MOD13Q1) for the Upper Blue Nile/Abay basin from 2001 to 2011.

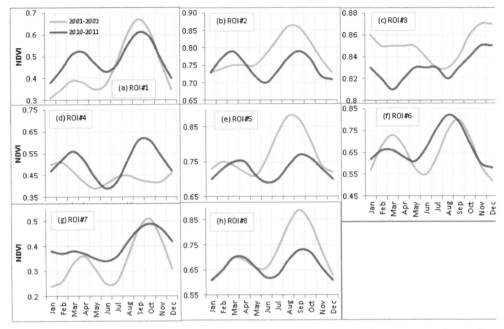

Figure 48 Fitted seasonal curves for 2001-2002 in green and 2010-2011 in red, derived from trends over the complete series. The intra-annual variations of generalized monthly NDVI over the first and last two-year periods of the time series (2001–2002 and 2010–2011) are shown for the 8 ROIs selected based on the amplitude composite images (Fig. 47). The spatial patterns of the colors exhibit significant trends in one or more of the amplitude shape parameters (i.e.A0, A1 and A2) and depict different land use and climate conditions

9.4.3   Trend break analysis

GIMMS-based trend break analysis in the inter-annual data series

The tread break analysis of ROIs shown in Figure 49 illustrates different trend behavior within a longer time series of satellite data. Trends are not always continuously increasing or decreasing but can change over time. For example, ROI #2 (Fig.49(b)) and ROI #4 (Fig.49(d)) represent gradual browning and greening trends, respectively, over the entire study period, while ROI #1 and ROI #3 describe abrupt browning and greening, respectively. ROI #1 characterized by trend breaks at December 1989 and August 1997 and ROI #3 shows a trend break at April 1998. Thus, the trend break analysis assisted identification of several change periods which might otherwise been overlooked through a commonly assumed fixed change trajectory analysis.

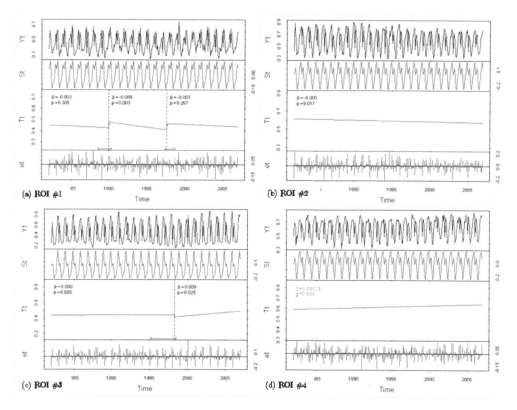

Figure 49 Examples of decomposition and trend break analysis of monthly GIMMS NDVI time series (1982-2006) for certain region of interests (ROIs) generated by BFAST approach a) ROI #1 (37.8E, 10.4N), b) ROI #2 (37.5E, 9.8N), c) ROI #3 (37.6E, 10.3N), d) ROI #4 (35.1E, 9.3N). The top panel in every plot shows the NDVI data, whereas the other three panels depict the individual components after decomposition. The seasonal (S_t) and remainder (e_t) components have zero mean while the trend component (T_t) shows the trend in NDVI. The slope coefficients (β) and the significance levels (P) at α value of 0.05 for each segment are given.

GIMMS-based trend break analysis in the HANTS seasonality parameters

Trend break analysis was performed to identify times when abrupt changes within the trends of harmonic series shape parameters occurred. The Pettitt test change point analysis for Class 1 shows that there is a significant shift ($k = 117$, $p = 0.001$) between two parts of the NDVI time series at the year 1998 (Fig. 50). The mean annual NDVI (A0) increased from a mean of

0.48 NDVI units observed for the period 1982-1998 to a mean value of 0.51 NDVI units observed for the period 1999-2006. A significant decreasing shift of Phase 1 occurred at the year 1997 for class 2 seasonal trend class (i.e. mainly woodland) from a mean phase angle of 217° observed during 1982-1997 to a mean phase angle of 210° during 1998-2006. This means the green-down period shifts to a later time of the year which has also implication on increasing length of growing season since 1998. Figure 50 also shows a significant increasing shift ($k = 124$, $p = 0.001$) of the annual amplitude from 0.15 to 0.17 NDVI units at the year 1998.

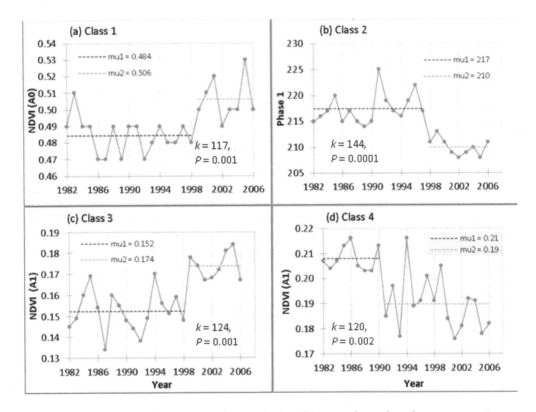

Figure 50 Pettitt test's change point analysis of harmonic series shape parameters for: (a) Class 1 (significant increases in Amplitude 0 (Mean NDVI)); (b) Class 2 (significant increases in Amplitude 0 and decrease in Phase 1 (the timing of annual peak greenness) together); (c) Class 3 (significant increases in Amplitude 1); and (d) Class 4 (significant decreases in Amplitude 1). The symbols mu1 and mu2 represent the mean values of the data series before and after the break-point, respectively.

MODIS-based trend break alanysis

The trend-break analysis did not indicate significant trend breaks using the BFAST approach in all of the sampled areas (Fig. 51) suggesting that vegetation activity in Abay basin showed a monotonic increasing or decreasing behavior, even if trend shifts are accounted for. Slowly acting processes such as land management practices or land degradation may cause such monotonic changes in the time series. Only ROI#4 shows a significant shift in the seasonal component at 2008 in Figure 51(c) and the break point indicates the start of double cropping in the area. However, the Pettitt's test identified different significant shifts at specific points in time for the three amplitude components. Table 48 shows the Pettitt's test result for identifying the time at which the shift occurs. The downward shift occurred during the period 2004-2006. For example, a highly significant ($p < 0.01$) decreasing shift in A0 occurred in 2004 for ROI#2 and in 2005 for ROI#5 and ROI#8. A highly significant ($p < 0.01$) decreasing shift in A1 occurred in 2006 for ROI#2 and 2005 for ROI#8. Thus, the significant decreasing shift in the vegetation condition of Abay basin occurred mainly during the period 2004-2006. This indicates a recent decline in vegetation activity. Significant increasing shift in the MODIS annual mean NDVI (A0) was observed in ROI#4 in 2007 ($p < 0.05$) and in ROI#7 in 2004 ($p < 0.01$) (Table 48). Annual amplitude also showed a significant increase in 2007 in ROI#4 ($p < 0.1$). Thus, 2007 can be considered as the time when a significant increase in vegetation condition occurred.

Table 48. Result of change point analysis of MODIS NDVI using Pettitt's test. The locations of Region of Interests (ROIs) are shown in Fig. 47. The numbers indicate the change point or year when significant change occurred. The positive (+) and the negative signs represent increasing and decreasing trend, respectively.

Variables	ROI#1	ROI#2	ROI#3	ROI#4	ROI#5	ROI#6	ROI#7	ROI#8
Annual mean NDVI (A0)	2003	2004	2004	2007	2005	2003	2004	2005
	*	***	**	(+)**	***		(+)***	***
Annual amplitude (A1)	2004	2006	2006	2007	2004	2006	2007	2005
	**	***		(+)*	**	*		***
Semi-annualamplitude A2)	2004	2004	2006	2005	2004	2003	2007	2007
				(+)*		*		**

$* = p < 0.1, *** = p < 0.05, *** = p < 0.001$

171

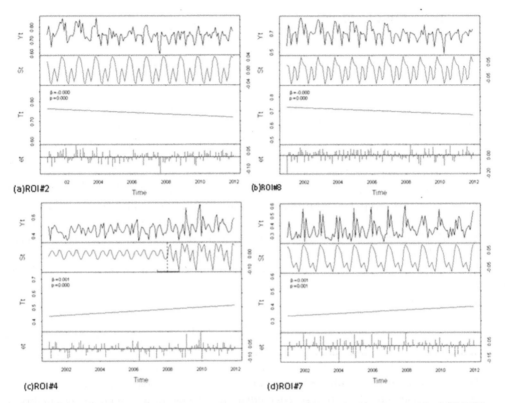

(a)ROI#2   (b)ROI#8   (c)ROI#4   (d)ROI#7

Figure 48 Examples of decomposition and trend break analysis of monthly MODIS NDVI time series (2001-2011) for certain region of interests (ROIs) for significant decreasing trends: (a) ROI #2 and (b) ROI #8 and for significant increasing trends: (c) ROI #4 and (d) ROI # generated by BFAST approach. The top panel in every plot shows the NDVI data, whereas the other three panels depict the individual components after decomposition. The seasonal (S_t) and remainder (e_t) components have zero mean while the trend component (T_t) shows the trend in NDVI. The slope coefficients (β) and the significance levels (P) at α value of 0.05 for each segment are given.

9.5 DISCUSSION

9.5.1 GIMMS-based inter-annual and seasonal trends

Vegetation trend analysis using the monthly GIMMS NDVI revealed a significant increase in the vegetation condition over 41.15% of the Abay basin for the period 1982-2006. Several studies have also reported widespread greening and increased vegetation growth and productivity for various areas of the world particularly in the Northern Hemisphere (Nemani et al., 2003; Slayback et al., 2003; Zhou et al., 2001). The north western and western parts of the basin that exhibited positive NDVI change are mainly covered with lowland vegetation such as woodland and shrubland. These areas are used for traditional uses such as shifting cultivation and livestock rearing, by the small indigenous population. One key characteristic of these areas is the periodic burning of vegetation, as it is evident from fire scares visible on Landsat images. The reason for this burning is not clear. Thus, the positive NDVI trend of these areas could be attributed to a decreasing trend of burning of vegetation. However, detailed investigation of these areas is crucial to be able to attribute the trend to particular cause.

The basin is characterized not only by significant inter-annual variation, but also by a significant trend in seasonality. More than half (59.52%) of land areas exhibited a significant trend in seasonality over the 25 year time period measured (1982-2006). Thus, a trend in seasonality is not a rare occurrence in Abay basin. Eastman et al. (2013) in their global study also concluded that a significant trend in seasonality is not a rare occurrence. Areas with significantly positive trend in mean annual NDVI (Amplitude 0) were found to be the dominant seasonal trend class, which accounts for 16.85% of all significantly trending areas and 10.03% of the total land area. The majority of this seasonal trend class occurs in areas of woodland and shrubland. A significant increase in mean annual NDVI without significant changes in any of the other harmonic shape parameters implies lengthening of the growing season. This result is consistent with a myriad of previous findings from both station observations and satellite observations. For example, studies in the Northern Hemisphere using phenology and temperature observations from stations have reported an increase in plant growth associated with a lengthening of the growing season. This result is consistent with previous findings from both station observations (Sparks et al., 2009) and satellite observations(Karlsen et al., 2007; Tucker et al., 2001; Zhou et al., 2001). A longer growing season could bring an increased photosynthesis by plants (Richardson et al., 2010) and it correlates strongly with annual gross primary productivity (Reed et al., 1994). Thus, it appears that woodland and shrubland areas of Abay basin are experiencing a consistent increase in productivity throughout the year.

The most likely driving factor of monotonic greening or browning could be changes in growth-limiting climatologies (Nemani et al., 2003), broad-scale land management practices, and persistent land use change driven by settlements (e.g., cropland expansion and urbanization) (Ramankutty et al., 2007). Strongest indication for greening in agricultural expansion areas such as Fincha and Belles sub-basins is most likely attributable to improved agricultural techniques. Abrupt browning could be caused by logging followed by regrowth or the already existed vegetation decline in some places might have amplified by to a period of persistently poor rains. However, abrupt greening could be caused by wet period induced by El Nino/La Nina Southern Oscillation warming events such as ENSO 1986/87, 1994/95, and 1997/98. The year 1997/1998 was identified as a trend break point using the Pettitt test. Thus, greening might represent recovery from drought or other disturbances such as forest fire (Anyamba and Tucker, 2005; Heumann et al., 2007; Olsson et al., 2005). Finally, although the GIMMS data sets have been thoroughly corrected (Tucker et al., 2005), AVHRR satellite platform changes could potentially cause trend breaks within the time series. The detected trend break time was compared with the time of AVHRR platform change (1985, 1988, 1994, 1995, 2000 and 2004) (Tucker et al., 2005) and it was found that that they are different. With regard to attribution of vegetation trend to the drivers, the coarser scale analysis only provides indications for more focused analysis, because the driving factors might act on a local scale much smaller than the resolution of GIMMS NDVI (i.e. 8 km). Thus, based on the information obtained from this coarser scale analysis area specific attribution of the increase in greenness to its causes is required for the Abay basin for the future.

9.5.2 MODIS-based inter-annual and seasonal trends

The vegetation trend analysis with a spatial resolution of 250 m MODIS (2001-2011) depicted a different perspective from that of 8km spatial resolution of GIMMS data set (1982-2006). About 36% of the UBN/Abay basin showed a significantly browning trend and only 1.19% of the basin showed a significantly greening trend over the period 2001-2011. The majority of the browning trend comes from Dabus, Belles, Wonbera and Didessa. The MODIS-based vegetation seasonal trend analysis during the period 2001-2011 revealed three conspicuous vegetation changes: (1) a change in the greenness pattern from a strong mean annual and annual cycle to strong semi-annual vegetation cycle, indicating significant changes from shrubland/woodland vegetation to double cropping (irrigation) (e.g. ROI#2, ROI#5 and ROI#8); (2) degradation of vegetation without complete removal (e.g. ROI#3); and (3) a change from a strong semi-annual vegetation cycle to strong mean annual NDVI, which could be related to increased vegetation activity in the area (e.g. on-farm eucalyptus plantation). These areas could have been masked if only the inter-annual vegetation trend analysis was performed. Thus, carrying out intra-annual trend analysis at medium resolution (250 m) would be helpful since most of the deforestations for the purpose of small scale irrigation are operating at local scale.

Although the BFAST approach of the trend break analysis indicated a monotonic increasing or decreasing vegetation condition without any significant trend breaks. However, based on Pettitt's test, it was possible to detect trend shifts for different HANTS shape parameters. The significant decreasing shift in the vegetation condition of Abay basin occurred mainly during the period 2004-2006. The greening tend observed from GIMMS data set might have been reversed into browning trend suggesting a localized decline in vegetation activity during the period 2004-2006. This result is consistent with the findings of de Jong et al. (2013b), who explained a trend reversal into browning trend not confined to single geographical region, but extended across all continents. In the 1980s and 1990s the Northern Hemisphere was found to become greener (Nemani et al., 2003;Zhou et al., 2001). The impact of sensor degradation on trends in MODIS NDVI (Wang et al., 2012) could be one cause for recent browning trend in the basin. Some anthropogenic factors could also be attributed to the observed recent browning trend in the basin. One of the possible drivers of browning in the northwest part of the basin especially in Rahad and some part Dinder sub-basins could be intra-regional resettlement (i.e. people from the same area with kin relations in the same locality) and its associated deforestation for cropland expansion (Dixon and Wood, 2007; Lemenih et al., 2014). The Amhara regional government of Ethiopia initiated 'voluntary' resettlement scheme since 2003 for the most chronically food insecure people from all zones of the region to potentially more productive, fertile and less populated parts such as Metema, Quara, Tach Armacheho and Tegede woredas of North Gondar administrative zone (NCFSE, 2003). In the Belles sub-basin land preparation for small to medium scale irrigation schemes could initially cause a decline in vegetation trend, but after a while the browning trend might be reversed to greening trend. Currently, private and government agricultural investments are underway in this sub-basin. Thus, the decreasing trend of vegetation greenness in the Belles sub-basin can be largely explained by the de-vegetation during land preparation for sugarcane plantation particularly in Alefa and Jawi weredas from Amhara regional state and Pawe and Dandur weredas from Benshagul Gumz regional states of Ethiopia. Such big croplands could contribute to the browning in the initial stage and to the greening once the cropland is established.

9.6 CONCLUSIONS

There are two notable features of long-term trend analysis conducted on monthly GIMMS NDVI data from 1982 to 2006. First, based on robust trend estimators (Theil-Sen slope) most part of the Upper Blue Nile basin (~77%) showed a positive trend in monthly NDVI with a mean rate of 0.0015 NDVI units (3.77% yr^{-1}), out of which 41.15% of the basin area depicted significant increases (p < 0.05) with a mean rate of 0.0023 NDVI units (5.59% yr^{-1}). Second, the upward trend (positive change) in NDVI is most intense in the northwestern and downward trend (negative change) in NDVI is intense in the southern and eastern part of the basin. Areas showing high NDVI variability were found to be the ones exhibiting long-term positive trends. The MODIS-based vegetation trend analysis revealed that about 36% of the UBN/Abay basin shows a significantly browning trend ($p < 0.05$) over the period 2001-2011 at an average rate of 0.0768 NDVI yr^{-1} using MODIS 250m spatial resolution. The majority of this browning trend comes from the sub-basins Dabus, Belles, Wonbera, and Didessa. The vegetation activity in the Belles sub basin has decreased at a faster rate (0.082NDVI yr^{-1}) than the other sub-basins. The greening trend observed from the GIMMS data set might have been reversed into browning trend and suggests a localized decline in vegetation activity in recent years. Anthropogenic factors such as resettlement and agricultural expansion could be attributed to the observed greening-to-browning reversal of vegetation conditions in the UBN basin.

Intra-annual trend analysis identified changes in vegetation condition that could have been masked if only the inter-annual vegetation trend analysis was performed. Changes in seasonality was found to be prevalent in the landscape of Upper Blue Nile/Abay basin as more than half (59.52%) of land areas exhibit a significant trend in seasonality over the 25 year time period measured (1982-2006). Only five types of seasonal trends were dominant out of which the largest proportion describes areas that are experiencing a uniform increase in NDVI throughout the year. It appears that woodland and shrubland areas of the basin experienced increase in vegetation productivity resulting from longer growing season. The seasonal trend analysis of MODIS NDVI (2001-2011) also confirmed that seasonal trend is not a rare occurrence in the Abay basin as 96% of the basin represents significant seasonal trends. Moreover, it revealed three conspicuous vegetation changes: (1) significant changes from shrubland/woodland vegetation to double cropping (irrigation); (2) degradation of vegetation without complete removal; and (3) localized increased vegetation activity.

Trend break analysis of GIMMS NDVI (1982-2006) showed that trends in vegetation conditions of the UBN basin were found to be not only monotonic, but also abrupt. The step change was detected in 1997/1998 in both inter-annual and intra-annual vegetation time series analysis. Thus, the trend break analysis assisted identification of several change periods which might otherwise be overlooked through a commonly assumed fixed change trajectory analysis. On the other hand, the trend break analysis performed on MODIS NDVI (2001-2011) using BFAST approach were not able to detect trend breaks suggesting that vegetation activity in Abay basin showed a monotonic increasing or decreasing behavior. Slowly acting processes such as land management practices or land degradation may cause such monotonic changes in the time series. However, the Pettitt test identified the significant decreasing shift in the vegetation condition of the Abay basin occurred mainly during the period 2004-2006. This indicates a recent decline in vegetation activity.

This study concludes that integrated analysis of inter-annual & intra-annual trend based on NDVI from GIMMS and MODIS revealed the following points:

- The greening trend of vegetation condition was followed by browning trend since mid-2000s in the Abay basin. Thus, there is an urgent need in increasing the vegetation activities in the basin.
- Intra-annual trend analysis was found to be very useful in identifying changes in vegetation condition that could have been masked if only the inter-annual vegetation trend analysis was performed.
- The MODIS-based intra-annual trend analysis revealed trends that were more linked to human activities;

177

Chapter 10

CLIMATIC CONTROLS OF NET PRIMARY

PRODUCTION AND WATER-USE EFFICIENCY IN THE

UPPER BLUE NILE (ABAY) BASIN[8]

10.1 ABSTRACT

Time series analysis of net primary production (NPP, amount of atmospheric carbon fixed) and water-use efficiency (WUE, amount of carbon uptake per unit of water use) and correlation analysis are effective tools to study impacts of land cover changes and their response to climatic variations. The results of this study show that NPP increased significantly ($P<0.05$) over 16.85% of the basin area by 6.25% (39.75 g C m^{-2}) and decreased significantly over 10.11% of the basin area by 6.75% (42.5 g C m^{-2}) over 25 years between 1982 and 2006. WUE increased significantly ($P<0.05$) by 14.88%, 0.13 g C kg^{-1}H$_2$O and decreased significantly ($P<0.05$) by 11.90%, 0.10 g C kg^{-1}H$_2$O over 24 years between 1983 and 2006. Woodland forest showed higher WUE than grassland and cropland suggesting that plantation of woodland-type of vegetation with higher WUE are required for the efficient management of carbon and water resources in the basin. The dominant climatic controlling factors of NPP vary according to aridity/humidity classes as follows: rainfall and temperature in humid zones; temperature and vapour pressure deficit (VPD) in semi-arid zones; and cloudiness in dry sub-humid zones of the basin. In the dry sub-humid and humid zones of the basin, WUE was correlated significantly and positively with maximum temperature, potential evaporation, and VPD, although no single climatic factor was correlated with WUE in semi-arid zones of the basin. The study concluded that any watershed activities designed for sustainable management of carbon and water

---

[8] *This chapter is based on* Teferi E. (2015) Climatic controls of net primary production and water-use efficiency in the Upper Blue Nile (Abay) basin (in preparation)

resources must consider the spatio-temporal patterns and controls of NPP and WUE across land cover types and A/H classes.

10.2 INTRODUCTION

It has been predicted that global climate changes will affect precipitation distributions, with increased drought conditions and other extreme climate events considerably by the end of the 21st (IPCC, 2007; Salinger, 2005). In the meantime, changes in other factors such as land use and land cover and atmospheric composition such as CO_2 interact with global climate change to influence both ecosystem carbon (C) and water budget of different terrestrial ecosystems (Niu et al., 2008; Yu et al., 2008). There is a considerable research interest in understanding the interactions between the carbon and water cycles (Davies et al., 2002; Ehleringer et al., 2000; Jackson et al., 2005) as it has been recognized as one of the major gaps in global change studies.

Water-use efficiency (WUE) that describes the ability of the vegetation to photosynthetically fix carbon per unit of water transpired, has been recognized as an important concept to study the interactions between the water and carbon cycles (Beer et al., 2009; Schwinning and Sala, 2004). Thus, WUE is an important index in climate change research and hydrological studies, as it reflects how the carbon and water cycles are linked. Basically, WUE relates primary production and total evaporation and more conveniently, it can be defined as the total evaporation costs of converting radiant energy to plant biomass. The rate at which radiant energy is converted to plant biomass is termed primary productivity and the sum total of the converted energy without the energy lost during plant respiration is called net primary productivity (NPP). Since NPP is directly related to WUE, different researchers have estimated WUE at ecosystem level as the ratio of NPP to total evaporation (E) expressed as WUE = NPP/E (Hu et al., 2008; Law et al., 2002; Tian et al., 2010). NPP has been identified as a primary monitoring variable in ecohydrological, land management, and environmental policy issues as it is a measurable and quantifiable characteristic of the biosphere. NPP represents the greatest annual carbon flux from the atmosphere to the biosphere and it is the major driver of variability in atmospheric CO_2 concentration (Hall et al., 1995; Keeling et al., 1996; Niemeijer, 2002). Thus, monitoring the trends of key biosphere attributes such as NPP and documenting the identified degradation could trigger policy shifts in land management.

Different studies have indicated that changes in climatic variables could alter ecosystems WUE via changes in NPP (Luo et al., 2008; Norby and Luo, 2004; Tian et al., 2010). NPP is primarily controlled by the climatic variables such as precipitation, temperature, and radiation (Hoeppner and Dukes, 2012; Nemani et al., 2003; Rustad et al., 2001). However, the contribution of these factors to vegetation growth differs according to the location of the ecosystem. Several studies from the northern hemisphere of mid-latitude and high latitude areas suggested that multiple controlling factors (e.g. CO_2 fertilization due to temperature increase, nitrogen deposition, and forest regrowth) have caused increasing NPP trend during the period 1982-1999 (MYNENI et al., 1997; Nemani et al., 2003). In tropical areas, the NPP variation is largely influenced by solar radiation due to cloudiness (Garbulsky et al., 2010;Zhao and Running, 2010). In drier and colder ecosystems NPP variation is related to precipitation and mean annual temperature (Schuur, 2003). In the arid and semi-arid ecosystems, water availability controls NPP variation (Fensholt et al., 2013). However, little is known about the controls of NPP and the associated WUE of vegetation in the Upper Blue Nile basin. So, there is a pressing need for quantitative assessments of patterns and controls of NPP and WUE in this area.

This study was conducted to investigate the spatial and temporal patterns of NPP and WUE and their underlying controlling factors. Specifically, this study wills address two main research questions: (1) how have NPP and WUE changed during the period 1982 -2006 across the study area? and (2) what are the climate variables controlling the inter-annual variation in NPP and WUE?. Understanding changes in NPP and WUE and their biotic and abiotic controls will support an early warning of land degradation and be used to support policy development for food security.

10.3 Materials and Methods

10.3.1 Data Sets

Three sets of climate data are used in this study: (1) Global Precipitation Climatology Centre (GPCC) data set; (2) the Climatic Research Unit Time series version 3.21 (CRU TS3.21) data sets; and (3) High resolution (0.1°) gridded rainfall data. The monthly precipitation data developed by the Global Precipitation Climatology Centre (hereafter GPCC) is used (Becker et al., 2012). The GPCC Full Data Reanalysis Version 6 for the period 1982–2006 with a spatial resolution of 0.5° The CRU TS3.21 data sets of monthly average daily minimum temperature (TMN), monthly average daily maximum temperature (TMX), monthly average daily mean temperature (TMP), and potential evaporation (E_p) with a spatial resolution of 0.5 degree are used (Harris et al., 2014). Although there is a monthly precipitation data of 0.5 degree spatial resolution from CRU, the performance of GPCC data is better than that of CRU for the study area (Dinku et al., 2008). The method used by CRU to compute E_p is the FAO (Food and Agricultural Organization) Penmane-Monteith model (Allen et al., 1994;Ekström et al., 2007). Monthly means of vapour pressure deficit (VPD), the difference between saturation and actual vapour pressure, was computed using CRU data and the Tetens formula (Buck, 1981):

$$VPD\ (kPa) = \left\{\left[a * \exp\left(\tfrac{b*TMX}{TMX+c}\right) + a * \exp\left(\tfrac{b*TMN}{TMN+c}\right)\right]/2\right\} - a * \exp\left(\tfrac{b*T_{dew}}{T_{dew}+c}\right) \qquad [10.1]$$

where TMX is monthly average daily maximum temperature in °C *and TMN* is the monthly average daily minimum temperature in °C, T_{dew} is the dew point temperature estimated from TMN (New et al., 1999). The constants *a, b,* and *c* were set based on Shuttleworth et al (Shuttleworth and Maidment, 1992) as a = 0.6108 kPa, $b = 17.27$, and c = 237.3° C.

10.3.2 Estimation of Net Primary Productivity Using Satellite Data

Different remote sensing data-driven light use efficiency models such as Global Production Efficiency Model (GLO-PEM) (Prince and Goward, 1995), Carnegie-Ames-Stanford approach (CASA) model (Imhoff et al., 2004) and Moderate Resolution Imaging Spectro radiometer (MODIS) MOD17 algorithm (Heinsch et al., 2003; Running et al., 2004) have been developed to produce spatiotemporal pattern of NPP of the terrestrial biosphere over large areas. However, a long-term NPP time series is still unavailable. Therefore, the GIMMS-based annual sum NDVI (ΣNDVI) was used as a surrogate for NPP due to the effectiveness of ΣNDVI in detecting and quantifying changes in NPP (Prince, 1991). Several studies demonstrated that there is a near-linear relationship between NPP and ΣNDVI in tropical grassland, cropland and sparse woodland (Asrar et al., 1985; Fensholt et al., 2006; Tucker and Sellers, 1986). Box et al (Box et al., 1989) showed that NDVI seems most closely related to primary production (or productivity), both net and gross, with a predictive accuracy for annual NPP comparable to that of climate-based NPP models. In order to estimate NPP in terms of carbon, a simple scaling of annual ΣNDVI into NPP was made by regressing the ΣNDVI on NPP from the MODIS NPP product (MOD17A3) for the overlapping years of the GIMMS and MODIS data sets (2000–2006). Rescaling of annual ΣNDVI into GLO-PEM NPP was also made by regressing them for the overlapping periods (1982-2000). Then the best rescaled NPP was selected based on its high correlation with ET. This kind of rescaling of ΣNDVI to get NPP in a data scarce situation was applied by different researchers (e.g.Bai et al., 2008; Prince et al., 2009).

10.3.3 Ecosystem Water Use Efficiency Calculation

There have been several different forms of relationship used to characterize WUE, and these have been summarized by Tanner and Sinclair (Tanner and Sinclair, 1983). The term water use efficiency often used by researchers to indicate the amount of dry matter production per unit of water consumed (Begg and Turner, 1976; Kramer, 1983). WUE is described in mathematical form as

$$Water-use\ efficiency = \frac{Dry\ matter\ production\ (g\ C\ m^{-2})}{Water\ consumed\ in\ evapotranspiration\ (mm)} \quad\quad [10.2]$$

On an ecosystem scale, WUE of whole ecosystem most researchers commonly compute WUE as the ratio of the main ecosystem fluxes such as gross primary production (GPP), gross ecosystem production (GEP), or net primary (ecosystem) production (NPP, NEP) to the water consumed by total evaporation (Brümmer et al., 2012; Hu et al., 2008; Law et al., 2002). NPP and total evaporation are tightly coupled (Schimel et al., 1997; Tian et al., 2010) and together they can reflect the close interactions between water and ecosystem productivity. Thus, in this study NPP-based WUE was computed for an individual grid cell as annual NPP in the grid cell divided by annual total evaporation (E). GIMMS NDVI-based global monthly E data from 1983 to 2006 were obtained from ftp://ftp.ntsg.umt.edu/pub/data/global_monthly_ET/. Detailed comparison of this data set with site-level observations can be found in Zhang et al.(2010) and Zhang et al. (2009).

10.4 RESULTS AND DISCUSSION

10.4.1 Spatial and Temporal Variation of ΣNDVI

Fig. 52a, b, and c show the green biomass production trends over the period 1982-2006 for Abay (UBN) basin, determined for each pixel by the TS slope estimator. During the period 1982-2006, the trends increased across 62.92% of the basin at a rate of 0.15%/yr and decreased over 37.08% of the basin at a rate of 0.13%/yr. However, only about 17.98% and 11.24% of the basin showed a significantly increasing trend (P<0.05) at a rate of 0.2935%/yr and significantly decreasing trend at a rate of 0.2903%/yr (P<0.05), respectively (Fig. 52 (c)). Thus, sum NDVI (i.e. biomass production) increased significantly by 7.34% and decreased by 7.26% over the period 1982-2006. The rate of increase of ΣNDVI is almost equal to that of the decrease in the Abay (UBN) basin over the period 1982-2006.

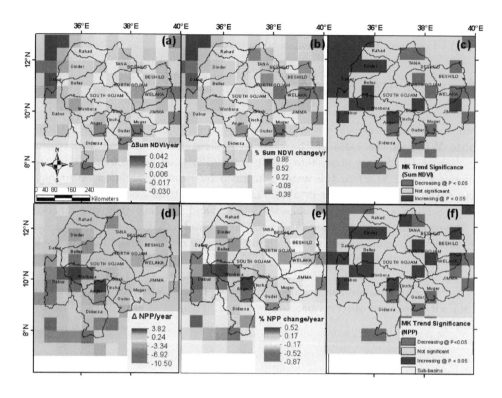

Figure 49 Trends in annual sum NDVI (ΣNDVI) and Net Primary Productivity (NPP) for the period 1982-2006: (a) median trend of ΣNDVI; (b) ΣNDVI trend in percentage; (c) Mann-Kendall (Shukla et al.) trend significance for ΣNDVI; (d) median trend of NPP; (e) NPP trend in percentage; (f) Mann-Kendall (MK trend significance) for NPP

10.4.2 Spatial and Temporal Variation of Net Primary Productivity

In order to select the best estimate of NPP for use in WUE computation, three different estimates of NPP were plotted against E. Since NPP and E are tightly coupled (Schimel et al., 1997;Tian et al., 2010), NPP with high correlation with ET was selected. Fig. 53 shows the scatter plot of the relationship between E and three different estimates of NPP. ΣNDVI rescaled on NPP from MOD17A3 ($NPP_{\Sigma NDVI\_MODIS}$) was selected due to its higher coefficient of determination ($r^2=0.64$).

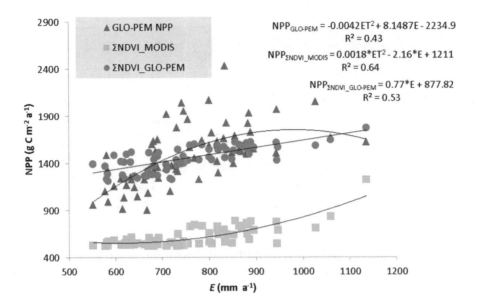

Figure 50 Scatter plot of the relationship between different multi-year (1983-2000) mean net primary productivity (NPP) estimates and multi-year (1983-2000) mean total evaporation (E). (1) multi-year (1983-2000) annual sum NDVI (ΣNDVI) rescaled on NPP from GLO-PEM (NPP$_{\Sigma NDVI\_GLO-PEM}$) versus E, (2) ΣNDVI rescaled on NPP from MOD17A3 (NPP$_{\Sigma NDVI\_MODIS}$) versus E, and (3) annual NPP from GLO-PEM (NPP$_{GLO-PEM}$) versus E.

Generally, the western and south-western part of the basin exhibited higher long-term mean annual NPP greater than 800 g C m^{-2} yr^{-1} (Fig. 54a) and higher annual NPP variability (Fig. 54b). Sub-basins such as Dabus, Wonbera, Dinder and Belles experienced increase in NPP at a rate about greater than 1 g C m^{-2} yr^{-1}. Based on the overlay analysis of the significantly increasing NPP pixels on land cover types, it was found that the observed significant increase in NPP was taking place on woodland land cover types. Fig. 54c Shows the absolute anomaly for annual NPP in 2004. High positive anomaly occures over the western and northeastern part of the basin and particularly in Wonbera sub-basin the NPP anomaly is even higher than 50 g Cm^{-2}yr^{-1}. Negative NPP anomalies are observed over Didessa, Muger, Guder, Fincha, South Gojjam and Lake Tana sub-basins.

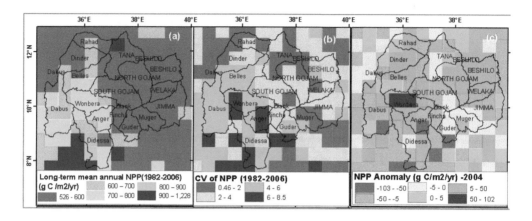

Figure 51 Inter-annual variability of NPP: (a) long-term mean annual NPP, (b) coefficient of variation (CV) of NPP, and (c) absolute NPP anomaly of the year 2004 for annual time step

During 1982-2006, it was found that the average annual NPP of the terrestrial ecosystems of Abay (UBN) basin ranges from 526 g C m^{-2}yr^{-1} to 1228 g C m^{-2}yr^{-1} and with spatial average of 624 g C m^{-2} yr^{-1}. During the period, about 65% of the basin showed a positive trend in NPP with a mean rate of 0.88 g C m^{-2} yr^{-1} (0.14% yr^{-1}) and the remaining 35% of the basin exhibited a decreasing trend with a mean rate of 0.81 g C m^{-2}yr^{-1} (0.13% yr^{-1}). NPP increased significantly ($P<0.05$) over 16.85% of the basin area, with a mean rate of 1.59 g C m^{-2} yr^{-1} (0.25% yr^{-1}) and decreased significantly over 10.11% of the basin, with a mean rate of 1.70 g C m^{-2} yr^{-1} (0.27% yr^{-1}). NPP increased significantly ($P<0.05$) by 6.25%, 39.75 g C m^{-2} over 25 years between 1982 and 2006. Globally, NPP increased by 6.17%, over 18 years, between 1982 and 1999 (Nemani et al., 2003). Thus, the increase in NPP of Abay basin is nearly equal to the global increase in NPP computed by Nemani et al (Nemani et al., 2003).

Table 48 Trends in NPP and WUE, and their climatic controls for the Upper Blue Nile/Abay basin during the period 1982-2006

Parameters	Trend estimator	Slope		MK-Z trends	
		Pos.	Neg.	Sig. Pos.	Sig. Neg.
Annual sum NDVI (ΣNDVI)	Δ ΣNDVI /yr	0.0093	0.0078	0.0176	0.0171
	% ΣNDVI change/yr	0.1517	0.1321	0.2935	0.2903
	Land area (%)	62.92	37.08	17.98	11.24
Net Primary Productivity (NPP)	Δ NPP/yr (g C m^{-2}yr^{-1})	0.88	0.81	1.59	1.70
	% NPP change/year	0.14	0.13	0.25	0.27
	Land area (%)	65.17	34.83	16.85	10.11
Water-use efficiency (WUE)	Δ WUE/yr (g C m^{-2} mm^{-1})	0.0021	0.0025	0.0052	0.0042
	% WUE change/year	0.2544	0.2964	0.6160	0.4956
	Land area (%)	47.19	52.81	5.62	14.61
Rainfall (P)	Δ rainfall (mm/yr) (TS Slope)	5.26	1.78	11.04	-
	% rainfall change/year	0.45	0.15	1.08	-
	Land area (%)	82	18	10.11	-
Maximum Temperature (TMX)	Δ TMX/yr (°C/yr) (TS Slope)	0.053	-	0.054	-
	% TMX change/year	0.20	-	0.20	-
	Land area (%)	100	-	97.75	-
Mean Temperature (TMP)	Δ TMP/yr (°C/yr) (TS Slope)	0.023	0.001	0.027	-
	% TMP change/year	0.12	0.003	0.14	-
	Land area (%)	97.75	2.25	67.42	-
Potential evaporation (Ep)	Δ Ep /yr (mm/yr) (TS Slope)	0.28	-	0.28	-
	% Ep change/year	0.26	-	0.27	-
	Land area (%)	100	-	98.88	-
Cloud cover (CLD)	Δ CLD/yr (%/yr) (TS Slope)	-	0.23	-	0.234
	% CLD change/yr	-	0.49	-	0.48
	Land area (%)	-	100	-	85.39
Vapour pressure deficit (VPD)	Δ VPD/yr (kPa/yr) (TS Slope)	0.0059	-	0.0059	-
	% VPD change/yr	0.59	-	0.59	-
	Land area (%)	100	-	100	-

10.4.3 Spatial and Temporal Variation of Water-Use Efficiency

During the period 1983-2006, it was found that the multi-year average annual WUE of the terrestrial ecosystems of Abay (UBN) basin ranges from 0.59 g C kg^{-1}H$_2$O to 1.13 g C kg^{-1}H$_2$O and with spatial average of 0.84 g C kg^{-1}H$_2$O. This implies that about 0.84 g C had been fixed in the terrestrial vegetation of UBN as net primary productivity by using 1 kg water. WUE values obtained in this study were within the range of values reported in literature. Tian et al. (2010) estimated the WUE of terrestrial ecosystem of Southern United States and obtained an average of 0.71 g C kg^{-1}H$_2$O during 1895-2007. Although the area is not similar the range of WUE values of this study is in the same order of magnitude with that of Tian et al. (2010). During the period, about 47% of the basin showed a positive trend in WUE with a mean rate of g C kg^{-1}H$_2$O (0.25% yr^{-1}) and the remaining 53% of the basin exhibited a decreasing trend with a mean rate of 0.0025 g C kg^{-1}H$_2$O (0.30% yr^{-1}) (Table 49 and Fig. 55a, b, and c). WUE increased significantly ($P<0.05$) over 5.62% of the basin area, with a mean rate of 0.0052 g C kg^{-1}H$_2$O (0.62% yr^{-1}) and decreased significantly over 14.61% of the basin, with a mean rate of 0.0042 g C kg^{-1}H$_2$O (0.50% yr^{-1}). The relative proportion of areas with a significant decrease in WUE is three times greater than areas with a significant increase in the UBN basin. These areas could be considered as locations where detailed local scale investigations are required. WUE increased significantly ($P<0.05$) by 14.88%, 0.13 g C kg^{-1}H$_2$O and decreased significantly ($P<0.05$) by 11.90%, 0.10 g C kg^{-1}H$_2$O over 24 years between 1983 and 2006. Higher WUE (>0.84 g C kg^{-1}H$_2$O) was observed in the eastern, north-western, western and south-western UBN (Fig. 55), while low values of WUE (<0.84 g C kg^{-1}H$_2$O) were obtained for the central and southern UBN.

Table 49 Result of one-way ANOVA test for mean difference in WUE among land cover types and aridity/humidity classes. Post Hoc Multiple comparisons using Least Significant Difference (LSD) test was performed. Only comparisons with significant mean difference are shown in the table

Sources of variation	Multiple comparison*	Mean Difference	Sig.
Land cover types	Shrubland (0.888) vs cropland (0.800)	0.0882	0.005
(ANOVA: F=6.87, Sig.=0.000)	Shrubland (0.888) vs grassland (0.769)	0.1191	0.037
	Woodland (0.893) vs cropland (0.800)	0.0935	0
	Woodland (0.893) vs grassland (0.769)	0.1244	0.021
Aridity/humidity classes	Semi-arid (0.954) vs dry sub-humid (0.860)	0.0939	0.002
(ANOVA: F=16.38, Sig.=0.000)	Semi-arid (0.954) vs humid (0.798)	0.1554	0
	Dry sub-humid (0.860) vs humid (0.798)	0.0615	0.008

*Mean WUE values are indicated in bracket

Based on the ANOVA results (Table 50), there is a statistically significant mean difference in WUE among land cover types (F=6.87, $P =0.000$) and aridity/humidity classes (F=16.38, $P =0.000$). Tian et al. (2010) also found that WUE varied with vegetation type. WUE of different land cover types showed a decreasing order of: Woodland > Shrubland > Cropland > Grassland in the UBN basin. In general, woodland forest had higher WUE than grassland and cropland suggesting that forest ecosystems use water more efficiently. The results are in agreement with that of Xiao et al. (2005) who reported that forest and cropland sites had higher annual fluxes than grassland sites. Thus, due to high carbon sequestration capacity per unit of water consumed, forest plantation could provide a significant carbon sink. WUE of aridity/humidity classes showed a decreasing order of: Semi-arid > Dry sub-humid> humid ecosystems.

187

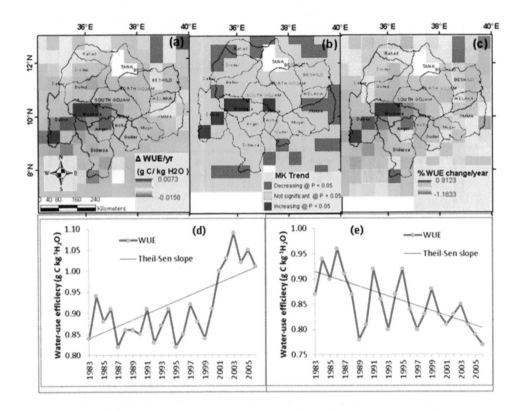

Figure 52 Inter-annual spatial and temporal pattern of water-use efficiency (WUE) of vegetation in the Upper Blue Nile (Abay) basin during the period 1983-2006. (a) median trend of WUE (b) Mann-Kendall trend significance, (c) percentage change of WUE, (d) the temporal pattern of increasing trend of WUE, and (e) the temporal pattern of decreasing trend pixels

10.4.4 Spatial and Temporal Variation of Climatic Variables

Cloud cover: Interannual variations and trends in cloud cover in relation to NPP were investigated because of their effects on solar radiation. The annual average cloud cover declined about 5.85% over 25 years (1982-2006) at a rate of 0.23% yr[-1] over 85.39% of the study area (Table 49). This result agrees with the results of different researchers worldwide (e.g.Eastman and Warren, 2013; Wylie et al., 2005), who also found a decreasing global trend in cloud cover. For instance, Eastman and Warren (2013) and Wylie et al. (2005) reported a 1.56% decline in global cloud cover over 39 years (1979-2009). Chen et al. (2002) also confirmed that due to the decline in cloud cover the thermal radiation emitted by Earth to space increased and reflected sunlight decreased in the tropics over the period 1985-2000. Wielicki et al. (2002) found that changes in both the annual average and seasonal tropical cloudiness are the major driver for the variability in tropical radiative energy budget (i.e. the long wave and short wave radiative fluxes). All the above studies agree closely with the results of this study.

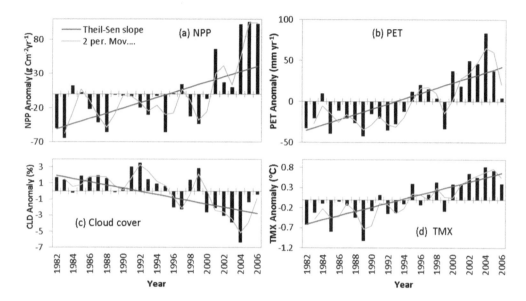

Figure 53 Interannual variations from 1982 to 2006 in net primary productivity (NPP) in relation to potential evaporation (E_p), cloud cover and maximum temperature at a location (35.75E 10.25N) in Abay/Upper Blue Nile basin. Bars indicate standardized anomalies of annual total NPP (a), annual mean E_p (b), and annual mean cloud cover (c). Green line indicates 2-year running means and red line indicates a long-term median trend (Theil-Sen slope)

Trends in cloud cover anomaly with respect to the 1982-2006 base period for sample pixel in the Abay basin is shown in Figure 56c. The anomaly of the annual mean cloud cover observed in 2004 was -6.38% and it was the first highest negative anomaly since 1982. The decreasing trend started since 1996 although there is a 2-year interruption. The positive and negative cloud cover anomalies are in line with the negative and positive anomalies of NPP, respectively.

Temperature: TMX, TMN, and TMP were computed to determine their relationship with NPP. Annual average maximum temperature significantly increased over about 98% of basin's land at a rate of 0.054°C yr^{-1}, indicating a 1.35°C (5%) increase in TMX during the period 1982-2006 (Table 49). There is a significant increase in mean temperature at annual time scale ($P < 0.05$) at a rate of $0.027°C$ yr^{-1} over about 67% the study area, indicating a $0.7°C$ (3.5%) increase in TMP during the study period. The increase in TMN (annually and seasonally) was found to be not significant ($P > 0.05$) over Abay basin. Irrespective of the method and scale used, several studies have also reported a rise in temperature in the study area. A recent study by Mengistu et al. (2014) reported a significant increase in maximum temperature ($0.01°C$ yr^{-1}) and minimum temperature ($0.015°C$ yr^{-1}) at annual time scale for Abay basin using reconstructed gridded data by ordinary least square method. Tekleab et al. (2013) also found significant increases in maximum, minimum and mean temperature at annual and seasonal timescale for more than half of the station used in their study for Abay basin. As opposed to this study, the above studies have found a significant increase in minimum temperature at annual time scale.

Rainfall: Based on the median slope computed upon GPCC v6 rainfall data during 1982-2006, the annual rainfall total shows a positive trend over 82% of the basin land area, out of which only 10.11% of the area shows a significantly increasing trend ($P<0.05$) at rate of 11 mm yr^{-1}(Table 49). The fact that the majority of the basin showed a non-significant increasing trend of annual rainfall is consistent with the findings of Mengistu et al. (2014). Mengistu et al. (2014) reported that rainfall in Abay basin showed a statistically non-significant increasing trend of 35mm per decade at the annual timescale. In contrast, Verdin et al. (2005) and Conway (2000) did not find any trend in rainfall over the north-eastern Ethiopian highlands.

Potential evaporation: Potential evaporation showed a significant increase (P < 0.05) at a rate of 0.28 mm yr^{-1} (2.7% per decade) over 98.88% of the basin during the period 1982-2006 (Table 49). Studies on trends of E_p in Ethiopia are very rare and it is very difficult to compare results. Regarding the global trend of PE, Barella-Ortiz et al. (2013) reported increasing trend of E_p estimated using PM-FAO method as 1.13% per decade. The trend results of PE are not consistent in literatures for different part of the world or even for the same place. Increasing trends in E_p were reported for the Mediterranean areas (e.g.Chaouche et al., 2010;Espadafor et al., 2011) or in the Middle East (Tabari et al., 2012; Tabari and Aghajanloo, 2013; Tabari and Talaee, 2011). While some other studies concluded that pan evaporation and potential evaporation have decreased in the last decades in Asia and North-America (Fan and Thomas, 2013;Hobbins et al., 2004;Roderick and Farquhar, 2004). Peterson et al. (1995) have presented decreases in pan evaporation over most of the USA and the former USSR between 1950 and 1990. Such decreases are generally not expected due to the observed increasing temperature and precipitation trends, resulting in an 'evaporation paradox' (Brutsaert and Parlange, 1998). Increased length of growing period observed in the northern hemisphere could be another evidence for overall increase in evaporation (White et al., 1999).

According to Matsoukas et al. (2011) the trends of potential evaporation appears to follow more closely the trends of energy availability than the trends of vapour transfer considerations. Availability of energy for evaporation is dependent directly on temperature and indirectly on cloud cover. As can be seen from Fig. 56b, the observed trend in Ep anomlaies is consistent with that of temperature and cloudiness. The Ep anomalies were almost constantly negative up to the year 1995 and the maximum negative anomaly (42 mm) was observed in 1989. The maximum positive anomaly was observed in 2004 (84mm).

Vapor pressure deficit (VPD): VPD showed a significant increase (P < 0.05) at a rate of 0.0059 kPa yr^{-1} or 14.75% over the 25 years during the period 1982-2006 in almost all pixels of the basin (Table 49). Allen et al. (1998) stated that VPD is the driving force that removes water vapor from the surface and thus, increases in VPD are indicative of decreased water availability. Steep increasing trend (0.058 kPa yr^{-1}) of global VPD was reported for the period 1983-2008 (Matsoukas et al., 2011). Although the direction of the trend of VPD is similar to that of the global trend, the rate of increase of VPD over the UBN is a bit lower than the global rate.

10.4.5 Relationship between NPP/WUE and Climate Variables

Table 51 clearly depicts that both NPP and WUE were found to vary significantly among land use and land cover (LULC) and aridity/humidity (A/H) classes. The interaction effect (LULC X A/H) was not significant suggesting that LULC and A/H independently affects changes in NPP or WUE.

Table 52 shows the Spearman's correlation coefficients of the relationship between NPP/WUE and climatic variables. Almost all climatic factors play significant roles in the trend and magnitude of NPP in humid areas. Thus, only one or two climatic factors could not be responsible for the trend. However, more strongly NPP is significantly correlated with rainfall ($r_s = 0.70$) and temperature ($r_s = 0.68$, $P < 0.01$) in the humid regions (Table 52). In semi-arid zones, NPP is correlated significantly more with TMX ($r_s = -0.74$, $P < 0.01$), Ep ($r_s = -0.73$, $P < 0.01$), and VPD ($r_s = -0.73$, $P < 0.01$) (Table 52). Ep by itself is a function of VPD and temperature, among other factors. Thus, VPD and TMX exhibit a dominant predictive linear relationship with NPP only in semi-arid areas. In dry sub-humid zones of the basin only cloud cover was found to be highly significantly correlated with NPP suggesting that solar radiation could play a significant role in vegetation growth in the zone. The dominant climatic controlling factors of NPP vary according to aridity/humidity classes as follows: rainfall and temperature in humid zones; temperature and VPD in semi-arid zones; and cloudiness in dry sub-humid zones of the basin.

Table 50 Univariate ANOVA test result for differences in NPP and WUE among land use and land cover types and aridity/humidity classes

Source of variation	Dependent Variable	F	Sig.
Land use and land cover (LULC)	NPP	4.77	0.004
	WUE	2.51	0.065
Aridity/Humidity (A/H)	NPP	3.26	0.044
	WUE	7.09	0.001
LULC X A/H	NPP	0.56	0.733
	WUE	0.60	0.697

Climatic controlling factors of WUE also vary based on A/H classes (Table 51). No single climatic factor correlated with WUE in semi-arid zones of the basin was observed (Table 52). In the dry sub-humid and humid zones of the basin, WUE was correlated significantly and positively with TMX (r_s = 0.40, $P < 0.05$), Ep (r_s = 0.48, $P < 0.05$), and VPD (r_s = 0.42, $P < 0.05$) (Table 52). In general Ep and its related factors exert the major control over WUE of the basin.

Table 51 The relationship between net primary productivity (NPP)/ecosystem water-use efficiency (WUE) and climatic variables using Spearman correlation coefficient (r_s).

r_s based on sources of variations	P	TMX	TMN	TMP	Ep	CLD	VPD
NPP							
Land use and land cover (LULC)							
All samples, n=89	0.59**	0.14	0.21	0.18	0.01	0.41**	0.02
Cropland	0.36*	0.2	0.27	0.26	0.14	0.41**	0.05
Shrubland	0.73**	-0.77**	-0.64*	-0.64*	-0.74**	-0.22	-0.78**
Woodland	0.49**	-0.64**	-0.65**	-0.63**	-0.69**	-0.35	-0.66**
Aridity/humidity (A/H)							
Semi-arid (0.2 < P/ Ep < 0.5)	0.52*	-0.73**	-0.71**	-0.72**	-0.74**	0.01	-0.73*
Sub-humid (0.5 < P/ Ep < 0.75)	0.27	0.34	0.35	0.36*	0.33	0.62**	0.34
Humid (P/ Ep > 0.75)	0.70**	0.64**	0.68**	0.68**	0.58**	0.62**	0.50**
WUE							
Land use and land cover (LULC)							
All samples, n=89	-0.29**	0.58**	0.56**	0.57**	0.64**	0.31**	0.57**
Cropland	-0.44**	0.36*	0.26	0.31*	0.61**	-0.43**	0.35*
Shrubland	-0.48	0.75**	0.78**	0.78**	0.77**	0.65*	0.76**
Woodland	-0.48*	0.44*	0.41*	0.44*	0.41*	0.53**	0.44*
Aridity/humidity (A/H)							
Semi-arid (0.2 < P/ Ep < 0.5)	-0.08	0.1	0.13	0.11	-0.02	0.38	0.09
Sub-humid (0.5 < P/ Ep < 0.75)	-0.27	0.40*	0.37	0.39*	0.48*	0.22	0.42*
Humid (P/ Ep > 0.75)	0.23	0.49**	0.46**	0.49**	0.57**	0.36*	0.44**

**. Correlation is significant at the 0.01 level (2-tailed).
*. Correlation is significant at the 0.05 level (2-tailed).

Climatic controls over NPP changes

NPP and cloud cover: When all samples are considered irrespective of land cover type, NPP was found to be correlated significantly and positively with cloudiness (r_s=0.41, P<0.01) and this positive correlation is because of the cropland ecosystem (Table 52). However, for shrubland (r_s=-0.22, P>0.05) and woodland (r_s=-0.35, P>0.05), non-significant negative correlations were found (Table 52). Several studies also have reported higher photosynthesis (i.e. NPP) on cloudy days compared to clear days (Alton et al., 2007; Law et al., 2002; Letts et al., 2005; Urban et al., 2007). The positive correlation between NPP with cloudiness could be explained by the reason that as cloudiness decreases (i.e. under sunny conditions) higher VPD may decrease stomatal conductance to CO_2 diffusion resulting in decrease in photosynthesis or NPP (Raveh et al., 2003; Urban et al., 2007). In contrast, Nemani et al. (2003) explained the increased tropical NPP in response to the increase in solar radiation resulting from declining cloudiness and indicated a negative correlation between NPP and cloudiness. The negative correlations observed in this study for shrubland and woodland lends support to the explanation given by Nemani et al. (2003) although the correlations are not significant. Thus, the observed increasing trend of NPP in the UBN basin cannot be sufficiently explained by the decreasing trend of cloudiness. One of the major reason for this could be clouds simultaneously affect a number of environmental variables. So, a more detailed research is required in this regard in the basin.

NPP and temperature: The relationship between the observed elevated temperature and NPP is indirect one, as they have no significant correlation between them when all samples are included in the correlation analysis (Table 52). The relationship between NPP and temperature depends upon the aridity /humidity classes or LULC types. For instance, maximum temperature has negative influence on NPP in shrubland (r_s=-0.77) and woodland (r_s=-0.64) (Table 52). However, no significant correlation between NPP and temperature for cropland is observed. Increase in temperature in semi-arid zones of the basin would cause a reduced NPP, but in humid zones temperature could be one of the reasons for increased NPP as it is positively and significantly correlated with NPP in humid zones (Table 52). Other similar studies (e.g.Lin et al., 2010; McMahon et al., 2010) have also shown that a rise in temperature could enhance vegetation growth. This increase in NPP could be explained by the fact that warming increases rates of microbial processes including litter decomposition and N mineralization, thereby increasing the availability of nutrients, and increasing plant productivity. Thus, increase in temperature has a positive feedback to the environment by increasing vegetation growth through increasing nutrient availability and lengthening the effective growing season particularly in humid zones of the basin. Furthermore, warming might also bring a shift in thermal limits to agricultural crops and certain vegetation types (Dunne et al., 2003) and it might bring new area of land into cultivation.

NPP and vapor pressure deficit: NPP is significantly and positively correlated with leaf-to-air VPD in the humid zone of the basin ($r_s = 0.50$, $P < 0.01$) indicating that increases in VPD induces increase in NPP in energy-limited environment (Table 52). However, in semi-arid zones of the basin, NPP is negatively correlated with VPD ($r_s = -0.73$, $P < 0.05$) (Table 52). Warmer temperature induces increased VPD which in turn causes higher soil evaporation and more rapid depletion of soil moisture and brings extended periods of stomatal closure and hence reduces NPP in water-limited environment (Eamus et al., 2013). Since most of the shrubland and woodland are located in the semi-arid areas of the basin, they showed negative correlations (Table 52).

NPP and rainfall: NPP of shrubland ($r_s = 0.73$, $P < 0.01$) and Woodland ($r_s = 0.49$, $P < 0.01$) were found to be significantly correlated with rainfall (Table 52). In humid zones, rainfall appears to have a stronger functional relationship with NPP, as they are correlated significantly ($r_s = 0.70$, $P < 0.01$). The correlation is not significant in sub-humid zone of the basin ($r_s = 0.26$, $P > 0.05$). In semi-arid zones, NPP and rainfall are correlated marginally and significantly ($r_s=0.52$, $P<0.05$). Thus, rainfall could be considered as important vegetation growth factor particularly in humid zones and semi-arid zones of the basin.

NPP and potential evaporation: NPP was negatively and strongly correlated with Ep over shrubland ($r_s=-0.74$) and woodland ($r_s=-0.69$) indicating that increasing Ep trends would lead to decreased NPP especially, in semi-arid zones of the basin ($r_s = -0.74$, $P < 0.01$) (Table 52). However, in humid zones, Ep increases linearly with NPP ($r_s = 0.58$, $P < 0.01$). Thus, the observed increase in NPP in this study could be attributed to the increase in Ep only for humid zones of the basin.

Climatic controls over ecosystem WUE

WUE and Temperature: Air temperature is positively and significantly correlated with WUE ($r_s=0.58$ (TMX), $r_s=0.56$ (TMN) and $r_s=0.57$ (TMP), $P < 0.01$) (Table 52). The increasing temperature trend reported by this study and other studies (e.g.Mengistu et al., 2014; Tekleab et al., 2013) could have a significant impact in increasing the ecosystem WUE in the UBN (Abay) basin. Increasing temperature (Climate warming) is normally associated with increased atmospheric CO_2 concentrations (IPCC, 2007). NPP and WUE can be increased because of the decline in stomatal conductance in response of vegetation to a warmer, CO_2-enriched environment (Ferretti et al., 2003; Nelson et al., 2004). This effect is sometimes referred to as the direct 'fertilization effect' (Beedlow et al., 2004). So, land use changes such as deforestation and wetland conversion to agriculture could induce CO_2 fertilization effect in the basin.

WUE and rainfall: WUE is significantly and negatively correlated with P ($r_s=-0.29$, $P < 0.01$) and indicating that ecosystems from lower-rainfall areas have higher water-use efficiency (Table 52). Soil moisture availability is one of the factors affecting stomatal aperture (Cowan, 1977; Schulze and Hall, 1982). For example, when a plant is subjected to water stress, stomata tend to close. This response is controlled by both electrical and hydraulic signals in response to soil moisture availability (Grams et al., 2007). Therefore, the result of the reduced stomatal conductance is increased WUE (Long et al., 2004). This is consistent with the correlation analysis of this study. In the UBN basin rainfall is significantly increasing ($P < 0.05$) in 10.11% of the basin area (Table 49) and thus, WUE would be decreasing in these areas.

WUE and cloudiness: When all samples are considered cloudiness is significantly and positively correlated with WUE ($r_s = 0.31$, $P < 0.01$) (Table 52). Consistent with this result, several studies have shown that cloudiness increased ecosystem WUE of shrubs and forest (e.g.Freedman et al., 2001; Min, 2005; Zhang et al., 2011a). This could be attributed to the changes in many other factors associated with clouds, such as temperature, moisture, latent heating, and precipitation, which may have direct and indirect influences on carbon assimilation and water use efficiency. In woodland ecosystem, cloudiness is relatively more strongly correlated with WUE as compared to other climatic variables. The observed decreasing trend of cloudiness in the tropical areas by this study and others (Chen et al., 2002; Wielicki et al., 2002) will have impact in decreasing the ecosystem WUE in these areas.

WUE and vapour pressure deficit (VPD): WUE is significantly and positively correlated with leaf-to-air VPD ($r_s = 0.57$, $P < 0.01$) (Table 52). A weak relationship was found between WUE and VPD over cropland ($r_s = 0.35$, $P < 0.05$) and woodland ($r_s = 0.44$, $P < 0.05$). However, in shrubland ecosystem a highly significant positive relationship between WUE and VPD was observed ($r_s = 0.76$, $P < 0.01$). The stomatal closure response to increasing VPD enables a plant to restrict excessive water loss before it develops severe water deficits and enhances WUE (Lange et al., 1971; Raveh et al., 2003; Urban et al., 2007).

WUE and potential evaporation: Ep shows a relatively stronger and significant control over WUE ($r_s = 0.64$, $P < 0.01$) than the other climatic factors considered in this study (Table 52). Penman's Ep itself is influenced by two factors: available energy for evaporation and leaf-to-air vapour pressure deficits (VPD). At most locations on the globe, Ep trends were found to be attributed primarily to changes in the radiation fluxes (i.e. available energy for evaporation indirectly inferred from by cloudiness and temperature) (Matsoukas et al., 2011), and secondarily to vapour transfer considerations. So, the observed decreasing trend of cloudiness and increasing trend of temperature in the UBN could cause a decline in stomatal conductance of leaves and this in turn increases WUE of vegetation. This response is highly significant in humid and shrubland and cropland areas (Table 52).

10.5 CONCLUSIONS

This study represents the first attempt to assess the patterns and controls of satellite-based estimates of NPP and WUE with the goal of understanding the interactions between carbon and water resources at basin scale. The results of this study show that NPP increased significantly ($P<0.05$) over 16.85% of the basin area by 6.25%, 39.75 g C m^{-2} and decreased significantly over 10.11% of the basin area by 6.75%, 42.5 g C m^{-2} over 25 years between 1982 and 2006. The proportion of area with significant increasing trend of NPP is greater than that of the decreasing trend, although the difference is quite small. WUE increased significantly ($P<0.05$) by 14.88%, 0.13 g C kg^{-1}H$_2$O and decreased significantly ($P<0.05$) by 11.90%, 0.10 g C kg^{-1}H$_2$O over 24 years between 1983 and 2006. The relative proportion of areas with a significant decrease in WUE is three times greater than areas with a significant increase in the UBN basin. In general, woodland forest had higher WUE than grassland and cropland suggesting that forest ecosystems use water more efficiently. Hence, management and plantation of woodland-type of vegetation with higher WUE are required for the efficient management of carbon and water resources in the basin. It was not possible to assess the impact of plantation forest on NPP and WUE because of the coarse spatial scale of the data used in this study. Certainly, the climatic controls of NPP and WUE need to be studied further using higher spatial resolution satellite imageries and field observation in future studies.

Quantifying trends of climatic variables for the period 1982-2006 is of particular interest because it provides insight to the observed trends of NPP and WUE. The annual average cloud cover declined about 5.85% over 25 years at a rate of 0.23% yr^{-1} over 85.39% of the study area. Declining clouds appear to be a major contributor to the observed increasing trend of warming their effects on solar radiation. TMX and TMP significantly increased at 0.054°C yr^{-1}and 0.027°C yr^{-1}, respectively. As the temperature increases, the evaporating power of the atmosphere also rises. Hence, VPD and E_p significantly increased at 0.0059 kPa yr^{-1} and 0.28 mm yr^{-1}, respectively. Based on GPCC v6 rainfall data, the annual rainfall total also showed a significantly increasing trend ($P<0.05$) at rate of 11 mm yr^{-1} over 10.11% of the UBN basin.

The climatic controlling factors of both NPP and WUE varied significantly among LULC and aridity/humidity classes. In humid areas of the basin, almost all climatic factors appear to play significant roles in the trend and magnitude of NPP, but NPP was significantly and positively correlated more strongly with rainfall and temperature. In semi-arid zones, NPP was correlated significantly and negatively more with TMX, E_p, and VPD. E_p by itself is a function of VPD and temperature, among other factors. Thus, VPD and TMX exhibit a dominant predictive linear relationship with NPP only in semi-arid areas. In dry sub-humid zones of the basin only cloud cover was found to be highly significantly correlated with NPP suggesting that solar radiation could play a significant role in vegetation growth in the zone. Climatic controlling factors of WUE also vary based on A/H classes. No single climatic factor correlated with WUE in semi-arid zones of the basin was observed. In the dry sub-humid and humid zones of the basin, WUE was correlated significantly and positively with TMX, Ep, and VPD. In general Ep and its related factors exert the major control over WUE of the basin. If the increasing temperature is combined with increased atmospheric CO_2 concentrations resulting from land use change (e.g. deforestation and conversion of wetlands), the increase in NPP and WUE can be amplified because of the decline in stomatal conductance in response of vegetation to a warmer, CO_2-enriched environment.

In an attempt to disentangle the climate-induced changes from human-induced changes in NPP using only high resolution (10km) rainfall gridded data, human-induced NPP changes (6.4%) were found to be the dominant in Abay basin than those driven by climate (3.5%). This pixel level identification can be further improved by including more climate variables such as temperature. Human-induced negative NPP changes were observed in Didessa sub-basin, which is hydrologically important to the UBN (Abay) basin. So, once this kind of rough but very crucial information is obtained, it can be used to give a focus to areas that demand detailed study and action on the ground. Furthermore, eco-hydrologic plots of water-versus-energy use efficiencies were also important in the identification of the period when the relative influence of human and/or climate-induced changes become dominant in the UBN basin, although it did not give the areal extent of the change.

Chapter 11

GENERAL CONCLUSIONS

11.1 MONITORING AND MODELLING CHANGES IN LAND USE AND LAND COVER

Identifying most systematic transitions in land use and land cover and its spatial statistical modelling through integrated use of remote sensing and geospatial technologies improved the identification, quantification, and understanding of determinants of most systematic transitions. The study found that the case study site (Jedeb watershed) has undergone significant land use/ cover alterations since 1957. A clear trend of growth in cultivated land until 1994 was found. The absence of a significant increase in the cultivated land since 1994 suggests that all the area that is suitable for cultivation has most likely been used already, or the land tenure system has effectively controlled spontaneous expansion of cultivation by local people. About 46% of the watershed area experienced a transition from one category to a different category of land use/cover over the 52 years considered. Out of the 46%, about 20% of the changed area was a net change while 25.9% was a swap change. Swap change is greater than net change suggesting the importance of the swapping component and common methods of land use/cover change study would miss this dynamic. Cultivated land and afro-alpine grassland tend to persist; riverine forest and shrubs and bushes tend to decrease; and grassland, woodland and marshland tend to gain or lose rather than persist. Identification of swapping change dynamics is important especially in the absence of the net change which may otherwise be interpreted as absence of change because the net change fails to capture the swapping component of the change.

The spatially explicit logistic regression model revealed that grassland-to-cultivated land and forest-to-grassland conversions are the most systematic transitions. In other words, more stable processes of change that evolve gradually in response to more permanent forces are acting on these conversions. Lower elevations, gentle slopes, less populated areas, locations near to markets/towns, and locations farther from roads increase the likelihood of grassland-to-cultivated land conversion in the Jedeb watershed. Farmers of the Jedeb watershed are inclined to convert grassland rather than natural forest for cultivation. The observed overall pattern has been conversion of natural woody vegetation to grassland and then grassland was subsequently converted to cultivated land. Therefore, the widely held view that expansion of agriculture is the primary cause for the loss of natural woody vegetation in Ethiopia was not found to hold true in the case of the Jedeb watershed. Forest edges, higher elevations, east-facing and steeper slopes, less populated areas, locations near roads and rivers, and locations with high soil wetness increase the likelihood of forest-to-grassland conversion. A 240 m-buffer of existing forest is likely to protect forest edges from conversion to grassland. Forests located near rivers and on wet soils (i.e. riverine forests) seem to have higher susceptibility to deforestation than the other forest types. Forests located near roads are more likely to be deforested, as lower travel costs increase financial benefits from forests. Thus, the loss of forest could be largely due to increased demand for wood for fuel, construction, farm implements and other uses. The rate of afforestation/reforestation (plantation) far outpaced that of deforestation in the periods 1972-1986 and 1986-1994, whereas recent (1994-2009) rates of deforestation of riverine forest (-4.44% a^{-1}) and shrubs and bushes (-3.69% a^{-1}) exceeded the rate of increase of recent (1994-2009) rate of plantation forest (0.88% a^{-1}). Thus, afforestation needs to be strengthened especially in the deforested areas of riverine vegetation and on the degraded shrubs and bushes. The spatial statistical modelling of most systematic transitions in this study suggests that spatial determinants of LULCC can be related to the well-established land change theories. The determinants of grassland-to-cultivated conversion model such as the travel time to markets reinforce evidence of von Thunen's model, whilst the geo-physical determinants of agricultural potential (elevation and slope) reinforce the Ricardian land change theory.

This thesis also demonstrated the use of remote sensing techniques to delineate headwater wetlands from non-wetlands and determine the dynamics over large areas such as the Choke Mountain range. Two major trajectories of wetland change were observed: (1) seasonal wetlands with low moisture to cultivated land, and (2) open water to bare land. In general, 607 km^2 of seasonal wetland with low moisture and 22.4 km^2 of open water were lost in the study area over the 20 years considered. This is an indication of future deterioration in wetland condition and calls for wetland conservation and rehabilitation through incorporating wetlands into watershed management plans so as to make wetlands continue providing their multiple functions which include flood and drought control, stream-flow moderation, groundwater recharge/discharge, sediment and nutrients detention, and pollutant retention. Further research on specific wetland spots is needed for accurate inventories of wetlands and to explore the reasons why and how wetlands are changing.

The landscape pattern analysis revealed that cultivated land constituted a matrix (i.e. > 50% of landscape) and likely to exert a dominant influence on the flora, fauna and many ecological processes in the watershed. This increase in dominance over time has caused several classes to occur in a clumped distribution and decreased spatial diversity of the watershed. More habitat loss and fragmentation observed in the 1986 landscape was associated with more irregularity in patch shapes (higher spatial heterogeneity). However, fragmentation of rangeland and marshland caused simplifications of patch shapes. Therefore, the idea that fragmentation has led to complexity of patch shapes is not a ubiquitous relationship. The observed decrease in ecologically important habitat extent (PLAND) of riverine forest, marshland, natural forest and rangeland appear to influence some major landscape functions such as soil erosion and water infiltration. Close monitoring of eco-hydrologically important landscape classes by land use planners is required for prompt mitigation of such adverse effects of landscape functions as a result of landscape pattern change. Thus, it is important to include landscape pattern analysis in sustainable planning of watershed management.

11.2 CHANGES IN SOIL HYDROLOGY IN RESPONSE TO CHANGES IN LAND USE AND LAND COVER

Soil degradation is one form of land degradation that is expected to come after LULCC which gives rise to changes in soil hydrological properties. The impact of LULC (land use and land cover) on soil organic matter content was found to be dependent on altitude, being higher in the upland part of the case study catchment (i.e. Jedeb). Additionally, the conversion of natural woody vegetation to cultivated land in the midland part decreased biomass productivity and reduced the quantity of biomass returned to the soil, and as a result the soil carbon decreased. The decline in soil organic carbon in the midland part of the watershed might bring adverse effects on soil structure. Therefore, conversion of degraded or marginal lands to restorative land uses should be adopted to minimize further depletion of soil carbon. This thesis showed that afforestation of barren land with *Eucalyptus* plantations can increase soil organic carbon content in the midland part. However, there was no evidence that this effect is true for the upland area as well. The conversions of natural woody vegetation to grassland and to barren land were accompanied by significant changes in soil bulk density. The compaction caused by cattle trampling in soils under grassland could be attributed to the significant higher bulk density. Therefore, one of the consequences of conversion of forest to grassland is soil compaction at least in the case of the Jedeb watershed. Thus, such types of unsustainable land use conversion need the attention of land users and land managers. The upland area and the midland area of the watershed have different degrees of soil compaction. Thus, a distinction should be made based on AEZs (Agro-Ecological Zones) during the selection of appropriate land management practices in the watershed. The midland part needs more efforts of reducing soil compaction than the upland part. Improved grazing land management has to be designed to minimize overgrazing, which can lead to soil compaction.

The impact of land cover change trajectories on soil moisture variability was evaluated using downscaled high-resolution soil moisture estimations through a synergistic use of microwave-optical/infrared data. Through comparisons to SMOS surface soil moisture, the downscaled multi-satellite surface soil moisture dataset (ECV SM) was demonstrated to provide a reasonably good estimate of soil moisture at a 1 km resolution at basin scale (i.e. UBN). Although such downscaling procedure provides some improvement, better methods that can downscale the ECV SM dataset to finer than 1 km spatial resolution are still needed. Both the downscaled and *in-situ* measured soil moisture were used to evaluate the impact of land cover change on soil moisture variation at the basin scale. Results indicated that there are significant differences in the mean soil moisture content between the reference trajectories (stable forest) and other trajectories considered as determined by the ANOVA result. The soils under stable forest trajectories had the highest soil moisture level. *In-situ* observations from the case study site, Jedeb watershed, also confirmed that the mean soil moisture levels under barren land and grassland significantly deviate from the mean soil moisture content of forest soils. Thus, removal of forests could cause a loss in the soil moisture storage capacity of soils. Significantly highest declines in soil moisture levels were observed due to conversion of shrubs and bushes to cropland and grassland. The observed mean difference in soil moisture between stable woodland and stable forest was small as compared to the mean difference of other trajectories. This means, if woodlands are protected from being deforested, it is possible to get nearly equivalent benefits of natural forest with regard to soil moisture management. Consequently, the study concludes that land cover change has a significant influence on the soil moisture regime of the basin. A better understanding of the relationships between soil moisture and land cover changes such as reforestation, afforestation, deforestation, and forest to grassland conversion is crucial for sustainable water resource development.

Measurements on the full range of soil moisture retention characteristics curve are often not available for use in hydrological models to assess land use impacts on soil hydraulic properties in data scarce areas such as the UBN basin. Moreover, the distinctive pedological properties of high altitude tropical soils of the UBN basin caused the development of new pedotransfer functions (PTFs). Readily available basic soil physical and chemical properties were found to predict the soil water retention characteristics with a mean root mean squared difference (MRMSD) of 0.0349 cm^3cm^{-3} and 0.0508 cm^3cm^{-3} for point PTFs and the continuous PTFs, respectively. The PTFs revealed that soil water retention depends not only on soil structure (i.e. bulk density and organic carbon) or soil texture but also on its chemical properties such as soil pH and CEC. Furthermore, collecting and integrating fragmented available soil survey data enables scientists and planners to generate new information such as maps of potential available water capacity at the basin scale. Future studies are recommended to be carried out on a database that is large enough to be representative of all or most parts of the textural triangle. The difference in compaction between topsoil and subsoil of medium textured soils resulted in a pronounced effect in the low-suction range of the soil water retention curve (SWRC). In contrast, in the very fine textured soil, the effect of compaction was pronounced at higher suctions. These results suggest that structure can markedly affect the soil water retention of medium textured soils, whereas the effect of soil structure is less pronounced in very fine textured soils. Since land use change affects soil structure by modifying properties such as soil organic matter and bulk density, the soil water retention capacity of medium textured soils can be enhanced by improving land use and management practices. However, further studies are required on impacts of land use management changes on soil hydraulic properties using PTFs at basin scale.

11.3 PATTERNS AND CLIMATIC CONTROLS OF VEGETATED LAND COVER

Apart from the broad timescale changes of vegetated land cover, understanding seasonal and inter-annual vegetation changes is important, as different eco-hydrological processes are operating at a variety of timescales. Most part of the Upper Blue Nile basin (~77%) showed a positive trend in monthly NDVI with a mean rate of 0.0015 NDVI units (~4% a^{-1}), out of which 41% of the basin area depicted significant increases (P < 0.05) with a mean rate of 0.0023 NDVI units (5.59% a^{-1}). The fine scale (250 m) MODIS-based vegetation trend analysis revealed that about 36% of the UBN basin shows a significantly browning trend ($P < 0.05$) over the period 2001-2011 at an average rate of 0.0768 NDVI a^{-1}. The greening trend observed from the GIMMS dataset might have been reversed into browning trend and suggests a localized decline in vegetation activity in recent years. Thus, there is an urgent need in increasing the vegetation activities in the basin.

Intra-annual trend analysis was found to be very useful in identifying changes in vegetation condition that could have been masked if only the inter-annual vegetation trend analysis was performed. Changes in seasonality was found to be prevalent in the landscape of the Upper Blue Nile basin as more than half (59.52%) of land areas exhibit a significant trend in seasonality over the 25 year time period measured (1982-2006). Only five types of seasonal trends were dominant out of which the largest proportion describes areas that are experiencing a uniform increase in NDVI throughout the year. The fine scale seasonal trend analysis of MODIS NDVI (2001-2011) revealed three conspicuous vegetation changes: (1) significant changes from shrubland/ woodland vegetation to double cropping (irrigation); (2) degradation of vegetation without complete removal; and (3) localized increase in vegetation activity more linked to human activities.

Land cover conversions and climate factors can significantly affect the ecology and hydrology of a landscape by modifying key ecosystem health indicators such as Water Use Efficiency (WUE) and Net Primary Productivity (NPP). This study represents the first attempt to assess the patterns and controls of satellite-based estimates of NPP and WUE with the goal of understanding the link between climate variables and carbon fixation and water use by vegetation. NPP increased significantly ($P <$ 0.05) over 16.85% of the basin area by 6.25%, 39.75 g C m^{-2} and decreased significantly over 10.11% of the basin area by 6.75%, 42.5 g C m^{-2} over 25 years between 1982 and 2006. The proportion of area with significant increasing trend of NPP is greater than that of the decreasing trend, although the difference is quite small. WUE increased significantly ($P < 0.05$) by 14.88%, 0.13 g C kg^{-1} H$_2$O and decreased significantly ($P < 0.05$) by 11.90%, 0.10 g C kg^{-1} H$_2$O over 24 years between 1983 and 2006. The relative proportion of areas with a significant decrease in WUE is three times greater than areas with a significant increase in the UBN basin. In general, woodlands had higher WUE than grassland and cropland suggesting that forest ecosystems use water more efficiently. Hence, management and plantation of woodland-type of vegetation with higher WUE will be required for an efficient management of carbon and water resources in the basin. It was not possible to assess the impact of plantation forest on NPP and WUE because of the spatial scale of the data used in this study. Certainly, the climatic controls of NPP and WUE need to be studied further using higher spatial resolution satellite imageries and field observation.

The climatic factors of both NPP and WUE varied significantly among LULC and aridity/humidity classes. In humid areas of the basin, almost all climatic factors appear to play significant roles in the trend and magnitude of NPP but more strongly NPP was significantly and positively correlated with rainfall and temperature. In semi-arid zones, NPP was correlated significantly and negatively more with maximum temperature (TMX), potential evaporation (Ep), and vapour pressure deficit (VPD). Ep by itself is a function of VPD and temperature, among other factors. Thus, VPD and TMX exhibit a dominant predictive linear relationship with NPP only in semi-arid areas. In dry sub-humid zones of the basin only cloud cover was found to be highly significantly correlated with NPP suggesting that solar radiation could play a significant role in vegetation growth in this zone. Climatic controlling factors of WUE also varied based on A/H classes. No single climatic factor correlated with WUE in semi-arid zones of the basin was observed. In the dry sub-humid and humid zones of the basin, WUE was correlated significantly and positively with TMX, Ep, and VPD. In general Ep and its related factors exert the major control over WUE of the basin. If the increasing temperature is combined with increased atmospheric CO$_2$ concentrations resulting from land use change (e.g. deforestation, conversion of wetlands, and burning of fuel), the increase in NPP and WUE can be amplified because of the decline in stomatal conductance in response of vegetation to a warmer, CO$_2$-enriched environment.

To conclude, human beings strongly depend on the sustainable availability of resources, such as food, water and energy. The continued supply of these resources can only be assured by sustainable land uses but these are easily threatened by inappropriate human activities. Human behaviour is intermingled with hydrological, biogeochemical, atmospheric and ecological processes through LULCC. This study provided a better understanding of soil hydrological impacts and climatic controls of LULCC. However, a range of key issues still needs to be better understood. Thus, future studies should include: (1) understanding the underlying socio-economic forces of LULCC that evolves from diverse human activities that are heterogeneous in their spatial, temporal and societal dimensions, (2) investigating the impacts of LULCC on climate through quantifying the partitioning of incoming solar radiation between evapotranspiration and sensible heat, (3) investigating the impact of LULCC on ecosystem water-use efficiency dynamics using higher spatial resolution satellite imageries and field observation, and (4) exploring the soil hydrological impacts of LULCC using both laboratory and field experiments, especially in areas where cropland-to-*Eucalyptus* conversions are common.

References

Aguilera, F., Valenzuela, L. M., and Botequilha-Leitão, A.: Landscape metrics in the analysis of urban land use patterns: A case study in a Spanish metropolitan area, Landscape and Urban Planning, 99, 226-238, 2011.

Allen, R., Smith, M., Pereira, L., and Perrier, A.: An update for the calculation of reference evapotranspiration, ICID bulletin, 43, 35-92, 1994.

Allen, R. G., Pereira, L. S., Raes, D., and Smith, M.: Crop evapotranspiration-Guidelines for computing crop water requirements-FAO Irrigation and drainage paper 56, FAO, Rome, 300, 1998.

Alo, C. A., and Pontius Jr, R. G.: Identifying systematic land-cover transitions using remote sensing and GIS: the fate of forests inside and outside protected areas of Southwestern Ghana, Environment and Planning B: Planning and Design, 35, 280-295, 2008.

Alton, P., North, P., and Los, S.: The impact of diffuse sunlight on canopy light-use efficiency, gross photosynthetic product and net ecosystem exchange in three forest biomes, Global Change Biology, 13, 776-787, 2007.

Amsalu, A., Stroosnijder, L., and Graaff, J. d.: Long-term dynamics in land resource use and the driving forces in the Beressa watershed, highlands of Ethiopia, Journal of Environmental Management, 83, 448-459, 2007.

Anderson, B. J. R., Hardy, E. E., Roach, J. T., and Witmer, R. E.: A Land Use And Land Cover Classification System For Use With Remote Sensor Data, Development, 2005, 28-28, 1976.

Angert, A., Biraud, S., Bonfils, C., Henning, C., Buermann, W., Pinzon, J., Tucker, C., and Fung, I.: Drier summers cancel out the CO2 uptake enhancement induced by warmer springs, Proceedings of the National Academy of Sciences of the United States of America, 102, 10823-10827, 2005.

Anyamba, A., and Tucker, C.: Analysis of Sahelian vegetation dynamics using NOAA-AVHRR NDVI data from 1981–2003, Journal of Arid Environments, 63, 596-614, 2005.

Asrar, G., Kanemasu, E., Jackson, R., and Pinter Jr, P.: Estimation of total above-ground phytomass production using remotely sensed data, Remote Sensing of Environment, 17, 211-220, 1985.

Bai, Z. G., Dent, D. L., Olsson, L., and Schaepman, M. E.: Proxy global assessment of land degradation, Soil use and management, 24, 223-234, 2008.

Baker, C., Lawrence, R., Montagne, C., and Patten, D.: Mapping wetlands and riparian areas using Landsat ETM+ imagery and decision-tree-based models, Wetlands, 26, 465-474, 2006.

Bala, G., Joshi, J., Chaturvedi, R. K., Gangamani, H. V., Hashimoto, H., and Nemani, R.: Trends and variability of AVHRR-derived NPP in India, Remote Sensing, 5, 810-829, 2013.

Baldocchi, D., Falge, E., Gu, L., Olson, R., Hollinger, D., Running, S., Anthoni, P., Bernhofer, C., Davis, K., and Evans, R.: FLUXNET: A new tool to study the temporal and spatial variability of ecosystem-scale carbon dioxide, water vapor, and energy flux densities, Bulletin of the American Meteorological Society, 82, 2415-2434, 2001.

Barella-Ortiz, A., Polcher, J., Tuzet, A., and Laval, K.: Potential evaporation estimation through an unstressed surface energy balance and its sensitivity to climate change, Hydrology and Earth System Sciences Discussions, 10, 8197-8231, 2013.

Bastiaanssen, W., Pelgrum, H., Soppe, R., Allen, R., Thoreson, B., and de C. Teixeira, A.: Thermal-infrared technology for local and regional scale irrigation analyses in horticultural systems, International Symposium on Irrigation of Horticultural Crops 792, 2006, 33-46, 2006.

BCEOM: Abbay River Basin Integrated Development Master Plan-Phase 2 – Water Resources – Hydrology, Ministry of Water Resources, Addis Ababa, 142 pp., 1998a.

BCEOM: Abbay River Basin Integrated Development Master Plan-Phase 2 – Land Resources Development – Reconnaissance Soils Survey, Ministry of Water Resources, Addis Ababa, 208 pp., 1998b.

BCEOM: Abbay River Basin Integrated Development Master Plan-Phase 2 – Land Resources Development – Land Cover/Land Use, Ministry of Water Resources, Addis Ababa, 83 pp.,1998c.

BCEOM: Abbay River Basin Integrated Development Master Plan-Phase 2 – Water Resources – Climatology, Ministry of Water Resources, Addis Ababa, 40 pp., 1998d.

BCEOM: Abbay River Basin Integrated Development Master Plan-Phase 2 – Natural Resources -Geology. Addis Ababa, Ethiopia, 98 pp., 1998e.

Becker, A., Finger, P., Meyer-Christoffer, A., Rudolf, B., Schamm, K., Schneider, U., and Ziese, M.: A description of the global land-surface precipitation data products of the Global Precipitation Climatology Centre with sample applications including centennial (trend) analysis from 1901–present, Earth System Science Data Discussions, 5, 921-998, 2012.

Beedlow, P. A., Tingey, D. T., Phillips, D. L., Hogsett, W. E., and Olszyk, D. M.: Rising atmospheric CO_2 and carbon sequestration in forests, Frontiers in Ecology and the Environment, 2, 315-322, 2004.

Beer, C., Ciais, P., Reichstein, M., Baldocchi, D., Law, B., Papale, D., Soussana, J. F., Ammann, C., Buchmann, N., and Frank, D.: Temporal and among-site variability of inherent water use efficiency at the ecosystem level, Global Biogeochemical Cycles, 23, 2009.

Begg, J. E., and Turner, N. C.: Crop water deficits, Adv. Agron, 28, 1976.

Beke, G. J., and MacCormick, M. I.: Predicting volumetric water retention for subsoil materials from Colchester County, Nova Scotia., Canadian Journal of Soil Science, 65, 233-236, 10.4141/cjss85-026, 1985.

Berberoglu, S., and Akin, A.: Assessing different remote sensing techniques to detect land use/cover changes in the eastern Mediterranean, International Journal of Applied Earth Observation and Geoinformation, 11, 46-53, 2009.

Bewket, W.: Land Cover Dynamics Since the 1950s in Chemoga Watershed, Blue Nile Basin, Ethiopia, Mountain Research and Development, 22, 263-269, 10.1659/0276-4741(2002)022[0263:lcdsti]2.0.co;2, 2002.

Bewket, W.: Household level tree planting and its implications for environmental management in the northwestern highlands of Ethiopia: a case study in the Chemoga watershed, Blue Nile basin, Land Degradation & Development, 14, 377-388, 2003.

Bewket, W., and Stroosnijder, L.: Effects of agroecological land use succession on soil properties in Chemoga watershed, Blue Nile basin, Ethiopia, Geoderma, 111, 85-98, 2003.

Bewket, W., and Teferi, E.: Assessment of soil erosion hazard and prioritization for treatment at the watershed level: case study in the Chemoga watershed, Blue Nile Basin, Ethiopia, Land Degradation & Development, 20, 609-622, 2009.

Bewket, W. and Sterk, G.: Dynamics in land cover and its effect on stream flow in the Chemoga watershed, Blue Nile basin, Ethiopia, Hydrol. Process., 19, 445–458, 2005.

Bhan, C.: Spatial analysis of potential soil erosion risks in Welo Region, Ethiopia: a geomorphological evaluation, Mountain Research and Development, 139-144, 1988.

Biazin, B., Stroosnijder, L., Temesgen, M., AbdulKedir, A., and Sterk, G.: The effect of long-term Maresha ploughing on soil physical properties in the Central Rift Valley of Ethiopia, Soil and tillage research, 111, 115-122, 2011.

Binford, M. W., and Buchenau, M. J.: Riparian greenways and water resources, Ecology of greenways, 69-104, 1993.

Biro, K., Pradhan, B., Buchroithner, M., and Makeschin, F.: Land use/land cover change analysis and its impact on soil properties in the northern part of Gadarif region, Sudan, Land Degradation & Development, 24, 90-102, 2013.

Bisrat, G.: Issues of Natural Resources Management in the Ethiopian Rangelands, Ethiopia's Experience in Conservation and Development, Addis Ababa, 1990,

Børgesen, C. D., Iversen, B. V., Jacobsen, O. H., and Schaap, M. G.: Pedotransfer functions estimating soil hydraulic properties using different soil parameters, Hydrological Processes, 22, 1630-1639, 2008.

Bosch, J. M., and Hewlett, J.: A review of catchment experiments to determine the effect of vegetation changes on water yield and evapotranspiration, Journal of hydrology, 55, 3-23, 1982.

Boserup, E.: The conditions of agricultural growth: the economics of agrarian change under population pressure, Aldine De Gruyter, New York, 1965.

Bot, A. J., Nachtergaele, F., and Young, A.: Land resource potential and constraints at regional and country levels, World Soil Resources Reports, 2000.

Bouma, J.: Using Soil Survey Data for Quantitative Land Evaluation, in: Advances in Soil Science, edited by: Stewart, B. A., Advances in Soil Science, Springer US, 177-213, 1989.

Bounoua, L., DeFries, R., Collatz, G. J., Sellers, P., and Khan, H.: Effects of Land Cover Conversion on Surface Climate, Climatic Change, 52, 29-64, 10.1023/a:1013051420309, 2002.

Bouyoucos, G. J.: Hydrometer method improved for making particle size analyses of soils, Agronomy Journal, 54, 464-465, 1962.

Box, E. O., Holben, B. N., and Kalb, V.: Accuracy of the AVHRR vegetation index as a predictor of biomass, primary productivity and net CO2 flux, Vegetatio, 80, 71-89, 1989.

Braimoh, A., and Vlek, P.: Land-Cover Change Trajectories in Northern Ghana, Environmental Management, 36, 356-373, 10.1007/s00267-004-0283-7, 2005.

Braimoh, A. K.: Random and systematic land-cover transitions in northern Ghana, Agriculture, Ecosystems & Environment, 113, 254-263, 10.1016/j.agee.2005.10.019, 2006.

Bray, R. H., and Kurtz, L.: Determination of total, organic, and available forms of phosphorus in soils, Soil science, 59, 39-46, 1945.

Bremner, J., and Mulvaney, C.: Nitrogen—total, Methods of soil analysis. Part 2. Chemical and microbiological properties, 595-624, 1982.

Bruand, A., Fernandez, P. P., and Duval, O.: Use of class pedotransfer functions based on texture and bulk density of clods to generate water retention curves, Soil Use and Management, 19, 232-242, 2003.

Bruijnzeel, L. A.: Hydrology of moist tropical forests and effects of conversion: a state of knowledge review, Hydrology of moist tropical forests and effects of conversion: a state of knowledge review., 1990.

Bruijnzeel, L. A.: Hydrological functions of tropical forests: not seeing the soil for the trees?, Agriculture, Ecosystems & Environment, 104, 185-228, 2004.

Brümmer, C., Black, T. A., Jassal, R. S., Grant, N. J., Spittlehouse, D. L., Chen, B., Nesic, Z., Amiro, B. D., Arain, M. A., and Barr, A. G.: How climate and vegetation type influence evapotranspiration and water use efficiency in Canadian forest, peatland and grassland ecosystems, Agricultural and Forest Meteorology, 153, 14-30, 2012.

Brutsaert, W., and Parlange, M.: Hydrologic cycle explains the evaporation paradox, Nature, 396, 30-30, 1998.

Buck, A. L.: New equations for computing vapor pressure and enhancement factor, Journal of Applied Meteorology, 20, 1527-1532, 1981.

Buckingham, E.: Studies on the movement of soil moisture, Bureau of Soils, US Department of Agriculture, Washington, DC, 1907.

Buenemann, M., Martius, C., Jones, J., Herrmann, S., Klein, D., Mulligan, M., Reed, M., Winslow, M., Washington-Allen, R., and Lal, R.: Integrative geospatial approaches for the comprehensive monitoring and assessment of land management sustainability: rationale, potentials, and characteristics, Land Degradation & Development, 22, 226-239, 2011.

Burt, T., and Butcher, D.: Topographic controls of soil moisture distributions, Journal of Soil Science, 36, 469-486, 1985.

Buytaert, W., Wyseure, G., De Bievre, B., and Deckers, J.: The effect of land-use changes on the hydrological behaviour of Histic Andosols in south Ecuador, Hydrological Processes, 19, 3985-3997, 2005.

Calder, I.: The hydrological impact of land-use change (with special reference to afforestation and deforestation), ODA Conference on Priorities for Water Resources Allocation and Management. Southampton, U. K., 1992, 91-102,

Calder, I. R.: Water-resource and land-use issues, Iwmi, 1998.

Calhoun, F., Hammond, L., and Caldwell, R.: Influence of particle size and organic matter on water retention in selected Florida soils, Proceedings–Soil and Crop Science Society of Florida, 1973, 111-113,

Calvet, J.-C., Wigneron, J.-P., Walker, J., Karbou, F., Chanzy, A., and Albergel, C.: Sensitivity of passive microwave observations to soil moisture and vegetation water content: L-band to W-band, Geoscience and Remote Sensing, IEEE Transactions on, 49, 1190-1199, 2011.

Campbell...1990

Campos, A. C., Etchevers, J. B., Oleschko, K. L., and Hidalgo, C. M.: Soil microbial biomass and nitrogen mineralization rates along an altitudinal gradient on the Cofre de Perote Volcano (Mexico): The importance of landscape position and land use, Land Degradation & Development, DOI: 10.1002/ldr.2185, 10.1002/ldr.2185, 2013.

Carlson, T. N., Gillies, R. R., and Perry, E. M.: A method to make use of thermal infrared temperature and NDVI measurements to infer surface soil water content and fractional vegetation cover, Remote Sensing Reviews, 9, 161-173, 1994.

Carlson, T. N., Gillies, R. R., and Schmugge, T. J.: An interpretation of methodologies for indirect measurement of soil water content, Agricultural and forest meteorology, 77, 191-205, 1995.

Celik, I.: Land-use effects on organic matter and physical properties of soil in a southern Mediterranean highland of Turkey, Soil and Tillage Research, 83, 270-277, 2005.

Chander, G., Markham, B. L., and Helder, D. L.: Summary of current radiometric calibration coefficients for Landsat MSS, TM, ETM+, and EO-1 ALI sensors, Remote sensing of environment, 113, 893-903, 2009.

Chaouche, K., Neppel, L., Dieulin, C., Pujol, N., Ladouche, B., Martin, E., Salas, D., and Caballero, Y.: Analyses of precipitation, temperature and evapotranspiration in a

French Mediterranean region in the context of climate change, Comptes Rendus Geoscience, 342, 234-243, 2010.

Chapman, H. D.: Cation-Exchange Capacity1, Methods of Soil Analysis. Part 2. Chemical and Microbiological Properties, agronomymonogra, 891-901, 10.2134/agronmonogr9.2.2ed.c6, 1965.

Chauhan, N., Miller, S., and Ardanuy, P.: Spaceborne soil moisture estimation at high resolution: a microwave-optical/IR synergistic approach, International Journal of Remote Sensing, 24, 4599-4622, 2003.

Chavez, P. S.: Image based atmospheric corrections - revisited and revised, Photogrammetric Engineering and Remote Sensing, 62, 1025-1036, 1996.

Chen, J., Carlson, B. E., and Del Genio, A. D.: Evidence for strengthening of the tropical general circulation in the 1990s, Science, 295, 838-841, 2002.

Choi, M., and Hur, Y.: A microwave-optical/infrared disaggregation for improving spatial representation of soil moisture using AMSR-E and MODIS products, Remote Sensing of Environment, 124, 259-269, 2012.

Chomitz, K. M., and Gray, D. A.: Roads, Land Use, and Deforestation: A Spatial Model Applied to Belize, The World Bank Economic Review, 10, 487-512, 1996.

Cohen, J.: A coefficient of agreement for nominal scales, Educational and psychological measurement, 20, 37-40, 1960.

Congalton, R. G., and Green, K.: Assessing the Accuracy of Remotely Sensed Data: Principles and Practices, CRC Press, 183-183 pp., 2009.

Conway, D.: The climate and hydrology of the Upper Blue Nile River, The Geographical Journal, 166, 49-62, 2000.

Cook, R. D.: Detection of Influential Observation in Linear Regression, Technometrics, 19, 15-18, 1977.

Coppin, P., Jonckheere, I., Nackaerts, K., Muys, B., and Lambin, E.: Review ArticleDigital change detection methods in ecosystem monitoring: a review, International Journal of Remote Sensing, 25, 1565-1596, 10.1080/0143116031000101675, 2004.

Cornelis, W. M., Ronsyn, J., Van Meirvenne, M., and Hartmann, R.: Evaluation of Pedotransfer Functions for Predicting the Soil Moisture Retention Curve, Soil Sci. Soc. Am. J., 65, 638-648, 10.2136/sssaj2001.653638x, 2001.

Coulson...1999

Cowan, I.: Stomatal behaviour and environment, Adv. Bot. Res., 4, 117-128, 1977.

Cushman, S. A., McGarigal, K., and Neel, M. C.: Parsimony in landscape metrics: strength, universality, and consistency, Ecological indicators, 8, 691-703, 2008.

Davies, W. J., Wilkinson, S., and Loveys, B.: Stomatal control by chemical signalling and the exploitation of this mechanism to increase water use efficiency in agriculture, New phytologist, 153, 449-460, 2002.

De Beurs, K., Wright, C., and Henebry, G.: Dual scale trend analysis for evaluating climatic and anthropogenic effects on the vegetated land surface in Russia and Kazakhstan, Environmental Research Letters, 4, 045012, 2009.

De Fraiture, C., Giordano, M., and Liao, Y.: Biofuels and implications for agricultural water use: blue impacts of green energy, Water Policy, 10, 67, 2008.

de Jong, R., de Bruin, S., de Wit, A., Schaepman, M. E., and Dent, D. L.: Analysis of monotonic greening and browning trends from global NDVI time-series, Remote Sensing of Environment, 115, 692-702, 2011.

de Jong, R., Verbesselt, J., Schaepman, M. E., and Bruin, S.: Trend changes in global greening and browning: contribution of short-term trends to longer-term change, Global Change Biology, 18, 642-655, 2012.

de Jong, R., Verbesselt, J., Zeileis, A., and Schaepman, M. E.: Shifts in global vegetation activity trends, Remote Sensing, 5, 1117-1133, 2013.

De Wit, A., and Su, B.: Deriving phenological indicators from SPOT-VGT data using the HANTS algorithm, 2nd international SPOT-VEGETATION user conference, 2005, 195-201,

DeFries, R., and Bounoua, L.: Consequences of land use change for ecosystem services: A future unlike the past, GeoJournal, 61, 345-351, 10.1007/s10708-004-5051-y, 2004.

DeFries, R., and Eshleman, K. N.: Land-use change and hydrologic processes: a major focus for the future, Hydrological Processes, 18, 2183-2186, 2004.

Dercon, S., and Hoddinott, J.: Livelihoods, growth, and links to market towns in 15 Ethiopian villages, 2005.

Dinku, T., Chidzambwa, S., Ceccato, P., Connor, S., and Ropelewski, C.: Validation of high-resolution satellite rainfall products over complex terrain, International Journal of Remote Sensing, 29, 4097-4110, 2008.

Dixon, A. B.: The hydrological impacts and sustainability of wetland drainage cultivation in Illubabor, Ethiopia, Land Degradation & Development, 13, 17-31, 2002.

Dixon, A. B., and Wood, A. P.: Wetland cultivation and hydrological management in eastern Africa: Matching community and hydrological needs through sustainable wetland use, Natural Resources Forum, 2003, 117-129,

Dixon, A. B., and Wood, A. P.: Local institutions for wetland management in Ethiopia: Sustainability and state intervention, Community-based water law and water resource management reform in developing countries, 130-145, 2007.

Dregne, H., Kassas, M., and Rozanov, B.: A new assessment of the world status of desertification, Desertification Control Bulletin, 20, 6-18, 1991.

Dunne, J. A., Harte, J., and Taylor, K. J.: Subalpine meadow flowering phenology responses to climate change: integrating experimental and gradient methods, Ecological Monographs, 73, 69-86, 2003.

Eamus, D., Boulain, N., Cleverly, J., and Breshears, D. D.: Global change-type drought-induced tree mortality: vapor pressure deficit is more important than temperature per se in causing decline in tree health, Ecology and evolution, 3, 2711-2729, 2013.

Eastman, J. R., Sangermano, F., Machado, E. A., Rogan, J., and Anyamba, A.: Global trends in seasonality of normalized difference vegetation index (NDVI), 1982–2011, Remote Sensing, 5, 4799-4818, 2013.

Eastman, R., and Warren, S. G.: A 39-yr survey of cloud changes from land stations worldwide 1971–2009: Long-term trends, relation to aerosols, and expansion of the tropical belt, Journal of Climate, 26, 1286-1303, 2013.

Ehleringer, J. R., Buchmann, N., and Flanagan, L. B.: Carbon isotope ratios in belowground carbon cycle processes, Ecological Applications, 10, 412-422, 2000.

Ehrenfeld, J. G.: Defining the limits of restoration: the need for realistic goals, Restoration ecology, 8, 2-9, 2000.

Ekström, M., Jones, P., Fowler, H., Lenderink, G., Buishand, T., and Conway, D.: Regional climate model data used within the SWURVE project–1: projected changes in seasonal patterns and estimation of PET, Hydrology and Earth System Sciences, 11, 1069-1083, 2007.

Espadafor, M., Lorite, I., Gavilán, P., and Berengena, J.: An analysis of the tendency of reference evapotranspiration estimates and other climate variables during the last 45 years in Southern Spain, Agricultural Water Management, 98, 1045-1061, 2011.

Fan, Z.-X., and Thomas, A.: Spatiotemporal variability of reference evapotranspiration and its contributing climatic factors in Yunnan Province, SW China, 1961–2004, Climatic change, 116, 309-325, 2013.

FAO: Assistance to Land-use Planning, Ethiopia: Land-use, Production Regions and Farming Systems Inventory. AG:DP/ETH/78/003, Technical Report 3., Rome, 1984.

Farley, K. A., Jobbágy, E. G., and Jackson, R. B.: Effects of afforestation on water yield: a global synthesis with implications for policy, Global change biology, 11, 1565-1576, 2005.

Feng, H., and Liu, Y.: Trajectory based detection of forest-change impacts on surface soil moisture at a basin scale [Poyang Lake Basin, China], Journal of Hydrology, 514, 337-346, 2014.

Fensholt, R., Sandholt, I., Rasmussen, M. S., Stisen, S., and Diouf, A.: Evaluation of satellite based primary production modelling in the semi-arid Sahel, Remote Sensing of Environment, 105, 173-188, 2006.

Fensholt, R., Rasmussen, K., Nielsen, T. T., and Mbow, C.: Evaluation of earth observation based long term vegetation trends—Intercomparing NDVI time series trend analysis consistency of Sahel from AVHRR GIMMS, Terra MODIS and SPOT VGT data, Remote Sensing of Environment, 113, 1886-1898, 2009.

Fensholt, R., Rasmussen, K., Kaspersen, P., Huber, S., Horion, S., and Swinnen, E.: Assessing land degradation/recovery in the African Sahel from long-term earth observation based primary productivity and precipitation relationships, Remote Sensing, 5, 664-686, 2013.

Fensholt, R., Langanke, T., Rasmussen, K., Reenberg, A., Prince, S. D., Tucker, C., Scholes, R. J., Le, Q. B., Bondeau, A., and Eastman, R.: Greenness in semi-arid areas across the globe 1981–2007—An Earth Observing Satellite based analysis of trends and drivers, Remote Sensing of Environment, 121, 144-158, 2012.

Fensholt, R., and Proud, S. R.: Evaluation of earth observation based global long term vegetation trends—Comparing GIMMS and MODIS global NDVI time series, Remote sensing of Environment, 119, 131-147, 2012.

Fernández, R. J., Wang, M., and Reynolds, J. F.: Do morphological changes mediate plant responses to water stress? A steady-state experiment with two C4 grasses, New Phytologist, 155, 79-88, 2002.

Ferreira, H., Botequilha-Leitão, A., Tress, B., Tress, G., Fry, G., and Opdam, P.: Integrating landscape and water-resources planning with focus on sustainability, Springer: Dordrecht, The Netherlands, 2006.

Ferretti, D., Pendall, E., Morgan, J., Nelson, J., Lecain, D., and Mosier, A.: Partitioning evapotranspiration fluxes from a Colorado grassland using stable isotopes: seasonal variations and ecosystem implications of elevated atmospheric CO2, Plant and Soil, 254, 291-303, 2003.

Field, A.: Discovering statistics using SPSS, Sage Publications, London, 2009.

Finlayson, C., Davidson, N., Spiers, A., and Stevenson, N.: Global wetland inventory–current status and future priorities, Marine and Freshwater Research, 50, 717-727, 1999.

Foody, G. M.: Status of land cover classification accuracy assessment, Remote Sensing of Environment, 80, 185-201, 10.1016/s0034-4257(01)00295-4, 2002.

Forman, R. T.: Some general principles of landscape and regional ecology, landscape ecology, 10, 133-142, 1995.

Forman, R., and Godron, M.: Landscape ecology. 619 pp, Jhon Wiley & Sons, New York, 1986.

Frazier, P. S., and Page, K. J.: Water body detection and delineation with Landsat TM data, Photogrammetric Engineering and Remote Sensing, 66, 1461-1468, 2000.

Freedman, J. M., Fitzjarrald, D. R., Moore, K. E., and Sakai, R. K.: Boundary layer clouds and vegetation-atmosphere feedbacks, Journal of Climate, 14, 180-197, 2001.

Friedl, M. A., Sulla-Menashe, D., Tan, B., Schneider, A., Ramankutty, N., Sibley, A., and Huang, X.: MODIS Collection 5 global land cover: Algorithm refinements and characterization of new datasets, Remote Sensing of Environment, 114, 168-182, 2010.

Friis, I., Demissew, S., and Breugel, P. v.: Atlas of the potential vegetation of Ethiopia, Det Kongelige Danske Videnskabernes Selskab, 2010.

Frohn, R. C., and Hao, Y.: Landscape metric performance in analyzing two decades of deforestation in the Amazon Basin of Rondonia, Brazil, Remote Sensing of Environment, 100, 237-251, 2006.

Fu, B., Chen, L., Ma, K., Zhou, H., and Wang, J.: The relationships between land use and soil conditions in the hilly area of the loess plateau in northern Shaanxi, China, Catena, 39, 69-78, 2000.

Furnival, G. M., and Wilson, R. W.: Regressions by leaps and bounds, Technometrics, 16, 499-511, 1974.

Gallo, K., Ji, L., Reed, B., Eidenshink, J., and Dwyer, J.: Multi-platform comparisons of MODIS and AVHRR normalized difference vegetation index data, Remote Sensing of Environment, 99, 221-231, 2005.

Gao, P., Mu, X.-M., Wang, F., and Li, R.: Changes in streamflow and sediment discharge and the response to human activities in the middle reaches of the Yellow River, Hydrology and Earth System Sciences, 15, 1-10, 2011.

Garbulsky, M. F., Peñuelas, J., Papale, D., Ardö, J., Goulden, M. L., Kiely, G., Richardson, A. D., Rotenberg, E., Veenendaal, E. M., and Filella, I.: Patterns and controls of the variability of radiation use efficiency and primary productivity across terrestrial ecosystems, Global Ecology and Biogeography, 19, 253-267, 2010.

Garten Jr, C. T., Post III, W., Hanson, P. J., and Cooper, L. W.: Forest soil carbon inventories and dynamics along an elevation gradient in the southern Appalachian Mountains, Biogeochemistry, 45, 115-145, 1999.

Gaskin, G., and Miller, J.: Measurement of soil water content using a simplified impedance measuring technique, Journal of agricultural engineering research, 63, 153-159, 1996.

Gautam, A. P., Webb, E. L., Shivakoti, G. P., and Zoebisch, M. A.: Land use dynamics and landscape change pattern in a mountain watershed in Nepal, Agriculture, ecosystems & environment, 99, 83-96, 2003.

Gebremedhin, B., and Swinton, S. M.: Investment in soil conservation in northern Ethiopia: the role of land tenure security and public programs, Agricultural Economics, 29, 69-84, 2003.

Gebremicael, T., Mohamed, Y., Betrie, G., van der Zaag, P., and Teferi, E.: Trend analysis of runoff and sediment fluxes in the Upper Blue Nile basin: A combined analysis of statistical tests, physically-based models and landuse maps, Journal of Hydrology, 482, 57-68, 2013.

Geist, H. J., and Lambin, E. F.: Proximate Causes and Underlying Driving Forces of Tropical Deforestation Tropical forests are disappearing as the result of many pressures, both local and regional, acting in various combinations in different geographical locations, BioScience, 52, 143-150, 2002.

Getahun, A.: Eucalyptus farming in Ethiopia: the case for eucalyptus woodlots in the Amhara region, 2002 Bahir Dar Conference Bahir Dar, 2002, 137-153, 2002.

Gilbert, R. O.: Statistical methods for environmental pollution monitoring, John Wiley & Sons, 1987.

Gillies, R. R., and Carlson, T. N.: Thermal remote sensing of surface soil water content with partial vegetation cover for incorporation into climate models, Journal of Applied Meteorology, 34, 745-756, 1995.

Gitau, M. W., Chaubey, I., Gbur, E., Pennington, J. H., and Gorham, B.: Impacts of land-use change and best management practice implementation in a Conservation Effects Assessment Project watershed: Northwest Arkansas, Journal of Soil and Water Conservation, 65, 353-368, 2010.

GLP: Science Plan and Implementation Strategy, IGBP Secretariat, Stockholm, 64, 2005.

Godfray, H. C. J., Beddington, J. R., Crute, I. R., Haddad, L., Lawrence, D., Muir, J. F., Pretty, J., Robinson, S., Thomas, S. M., and Toulmin, C.: Food security: the challenge of feeding 9 billion people, science, 327, 812-818, 2010.

Gómez-Plaza, A., Martınez-Mena, M., Albaladejo, J., and Castillo, V.: Factors regulating spatial distribution of soil water content in small semiarid catchments, Journal of hydrology, 253, 211-226, 2001.

Grams, T. E., Koziolek, C., Lautner, S., Matyssek, R., and Fromm, J.: Distinct roles of electric and hydraulic signals on the reaction of leaf gas exchange upon re-irrigation in Zea mays L, Plant, cell & environment, 30, 79-84, 2007.

Grayson, R. B., and Western, A. W.: Towards areal estimation of soil water content from point measurements: time and space stability of mean response, Journal of Hydrology, 207, 68-82, 1998.

Griffiths, R., Madritch, M., and Swanson, A.: The effects of topography on forest soil characteristics in the Oregon Cascade Mountains (USA): Implications for the effects of climate change on soil properties, Forest Ecology and Management, 257, 1-7, 2009.

Guerreiro, S. B., Kilsby, C. G., and Serinaldi, F.: Analysis of time variation of rainfall in transnational basins in Iberia: abrupt changes or trends?, International Journal of Climatology, 34, 114-133, 2014.

Hailu, A., Wood, A., and Dixon, A.: Interest groups, local knowledge and community management of wetland agriculture in South-West Ethiopia, International Journal of Ecology and Environmental Sciences, 29, 55-63, 2003.

Hale, S. R., and Rock, B. N.: Impact of Topographic Normalization on Land-Cover Classification Accuracy, Photogrammetric Engineering Remote Sensing, 69, 785-791, 2003.

Hall, D., Ojima, D., Parton, W., and Scurlock, J.: Response of temperate and tropical grasslands to CO2 and climate change, Journal of Biogeography, 537-547, 1995.

Han, T., Wulder, M., White, J., Coops, N., Alvarez, M., and Butson, C.: An efficient protocol to process Landsat images for change detection with tasselled cap transformation, Geoscience and Remote Sensing Letters, IEEE, 4, 147-151, 2007.

Hansen, A. J., DeFries, R. S., and Turner, W.: Land Use Change and Biodiversity: A Synthesis of Rates and Consequences during the Period of Satellite Imagery Land Change Science, in, edited by: Gutman, G., Janetos, A. C., Justice, C. O., Moran, E. F., Mustard, J. F., Rindfuss, R. R., Skole, D. L., Turner, B. L., and Cochrane, M. A., Remote Sensing and Digital Image Processing, Springer Netherlands, 277-299, 2004a.

Hansen, A., DeFries, R., and Turner, W.: Land Use Change and Biodiversity, in: Land Change Science, edited by: Gutman, G., Janetos, A., Justice, C., Moran, E., Mustard, J., Rindfuss, R., Skole, D., Turner, B., II, and Cochrane, M., Remote Sensing and Digital Image Processing, Springer Netherlands, 277-299, 2004b.

Harris, I., Jones, P., Osborn, T., and Lister, D.: Updated high-resolution grids of monthly climatic observations–the CRU TS3. 10 Dataset, International Journal of Climatology, 34, 623-642, 2014.

Harris, J. W., and Stöcker, H.: Handbook of mathematics and computational science, Springer, 1998.

Hartemink, A. E.: Soil fertility decline in the tropics with case studies on plantations, CAB International / ISRIC, Wallingford, 384 pp., 2003.

Hedberg, O.: Vegetation belts of the east african mountains, Svensk Botanisk Tidskrift, Sevensk Botanisk Tidskrrift 45, Svenska botaniska fôreningens, 63 pp., 1951.

Heinsch, F. A., Reeves, M., Votava, P., Kang, S., Milesi, C., Zhao, M., Glassy, J., Jolly, W. M., Loehman, R., and Bowker, C. F.: GPP and NPP (MOD17A2/A3) Products NASA MODIS Land Algorithm, 2003.

Helldén, U., and Tottrup, C.: Regional desertification: A global synthesis, Global and Planetary Change, 64, 169-176, 2008.

Henebry, G. M.: Global change: carbon in idle croplands, Nature, 457, 1089-1090, 2009.

Herold, M., Mayaux, P., Woodcock, C., Baccini, A., and Schmullius, C.: Some challenges in global land cover mapping: An assessment of agreement and accuracy in existing 1 km datasets, Remote Sensing of Environment, 112, 2538-2556, 2008.

Herrmann, S. M., Anyamba, A., and Tucker, C. J.: Recent trends in vegetation dynamics in the African Sahel and their relationship to climate, Global Environmental Change, 15, 394-404, 2005.

Heumann, B. W., Seaquist, J., Eklundh, L., and Jönsson, P.: AVHRR derived phenological change in the Sahel and Soudan, Africa, 1982–2005, Remote Sensing of Environment, 108, 385-392, 2007.

Hickler, T., Eklundh, L., Seaquist, J. W., Smith, B., Ardö, J., Olsson, L., Sykes, M. T., and Sjöström, M.: Precipitation controls Sahel greening trend, Geophysical Research Letters, 32, 2005.

Hill, J., and Sturm, B.: Radiometric correction of multitemporal Thematic Mapper data for use in agricultural land-cover classification and vegetation monitoring, International Journal of Remote Sensing, 12, 1471-1491, 1991.

Hillel, D.: Introduction to environmental soil physics, Introduction to environmental soil physics, 2004.

Hoaglin, D. C., Mosteller, F., and Tukey, J. W.: Understanding robust and exploratory data analysis, Wiley New York, 1983.

Hobbins, M. T., Ramírez, J. A., and Brown, T. C.: Trends in pan evaporation and actual evapotranspiration across the conterminous US: Paradoxical or complementary?, Geophysical Research Letters, 31, 2004.

Hocking, R. R.: A Biometrics invited paper. The analysis and selection of variables in linear regression, Biometrics, 32, 1-49, 1976.

Hodgson, M. E., and Shelley, B. M.: Removing the topographic effect in remotely sensed imagery, Erdas Monitor, 6, 4-6, 1994.

Hodnett, M. G., and Tomasella, J.: Marked differences between van Genuchten soil water-retention parameters for temperate and tropical soils: a new water-retention pedo-transfer functions developed for tropical soils, Geoderma, 108, 155-180, 2002.

Hoeppner, S. S., and Dukes, J. S.: Interactive responses of old-field plant growth and composition to warming and precipitation, Global Change Biology, 18, 1754-1768, 2012.

Holben, B. N.: Characteristics of maximum-value composite images from temporal AVHRR data, International Journal of Remote Sensing, 7, 1417-1434, 1986.

Holben, B., and Justice, C.: An examination of spectral band ratioing to reduce the topographic effect on remotely sensed data, International Journal of Remote Sensing, 2, 115-133, 1981.

Holsten, A., Vetter, T., Vohland, K., and Krysanova, V.: Impact of climate change on soil moisture dynamics in Brandenburg with a focus on nature conservation areas, Ecological Modelling, 220, 2076-2087, 2009.

Homaee, M., and Firouzi, A. F.: Deriving point and parametric pedotransfer functions of some gypsiferous soils, Soil Research, 46, 219-227, 2008.

214

Honnay, O., Piessens, K., Van Landuyt, W., Hermy, M., and Gulinck, H.: Satellite based land use and landscape complexity indices as predictors for regional plant species diversity, Landscape and urban planning, 63, 241-250, 2003.

Hosmer, D. W., and Lemeshow, S.: Applied Logistic Regression, Wiley Series in Probability and Statistics: Texts and References Section, John Wiley & Sons, New York, USA, 2000.

Hu, W., Shao, M., Han, F., and Reichardt, K.: Spatio-temporal variability behavior of land surface soil water content in shrub-and grass-land, Geoderma, 162, 260-272, 2011.

Hu, Z., Yu, G., Fu, Y., Sun, X., Li, Y., Shi, P., Wang, Y., and Zheng, Z.: Effects of vegetation control on ecosystem water use efficiency within and among four grassland ecosystems in China, Global Change Biology, 14, 1609-1619, 2008.

Huang, C., Wylie, B., Yang, L., Homer, C., and Zylstra, G.: Derivation of a tasselled cap transformation based on Landsat 7 at-satellite reflectance, International Journal of Remote Sensing, 23, 1741-1748, 2002.

Huete, A., Didan, K., Miura, T., Rodriguez, E. P., Gao, X., and Ferreira, L. G.: Overview of the radiometric and biophysical performance of the MODIS vegetation indices, Remote sensing of environment, 83, 195-213, 2002.

Hurni, H.: Degradation and conservation of the resources in the Ethiopian highlands, Mountain research and development, 123-130, 1988.

Hurni, H.: Principles of soil conservation for cultivated land, Soil Technology, 1, 101-116, 1988.

Hurni, H.: Degradation and conservation of soil resources in the Ethiopian Highlands, in: African Mountains and Highlands: Problems and Prospects, edited by: Messerli, B., and Hurni, H., Geographica Bernensia, African Mountains Association, Bern, Switzerland and Addis Abeba, Ethiopia, 1990.

Hurni, H., and Pimentel, D.: Land degradation, famine, and land resource scenarios in Ethiopia, World soil erosion and conservation., 27-61, 1993.

Hurni, H.: Agroecological Belts of Ethiopia, Switzerland/Ethiopia, 1999.

Hurni, H., Tato, K., and Zeleke, G.: The Implications of Changes in Population, Land Use, and Land Management for Surface Runoff in the Upper Nile Basin Area of Ethiopia, Mountain Research and Development, 25, 147-154, 10.1659/0276-4741(2005)025[0147:tiocip]2.0.co;2, 2005.

Imhoff, M. L., Bounoua, L., DeFries, R., Lawrence, W. T., Stutzer, D., Tucker, C. J., and Ricketts, T.: The consequences of urban land transformation on net primary productivity in the United States, Remote Sensing of Environment, 89, 434-443, 2004.

IPCC, I. P. O. C. C.: Climate change 2007: The physical science basis, Agenda, 6, 333, 2007.

Izenman, A. J.: Modern multivariate statistical techniques: regression, classification, and manifold learning, Springer, 2008.

Izenman, A. J.: Modern Multivariate Statistical Techniques: Regression, Classification, and Manifold Learning, Springer, 2009.

Jackson, R. B., Jobbágy, E. G., Avissar, R., Roy, S. B., Barrett, D. J., Cook, C. W., Farley, K. A., Le Maitre, D. C., McCarl, B. A., and Murray, B. C.: Trading water for carbon with biological carbon sequestration, Science, 310, 1944-1947, 2005.

Jagtap, S., Lall, U., Jones, J., Gijsman, A., and Ritchie, J.: Dynamic nearest-neighbor method for estimating soil water parameters, Transactions of the ASAE, 47, 1437-1444, 2004.

Jakubauskas, M. E., Legates, D. R., and Kastens, J. H.: Harmonic analysis of time-series AVHRR NDVI data, Photogrammetric Engineering and Remote Sensing, 67, 461-470, 2001.

Jarque, C. M., and Bera, A. K.: Efficient tests for normality, homoscedasticity and serial independence of regression residuals, Economics Letters, 6, 255-259, 1980.

Jensen, J. R., Narumalani, S., and Mackey, H.: Measurement of seasonal and yearly cattail and water lily changes using multi date SPOT panchromatic data, Photogrammetric Engineering and Remote Sensing, 59, 519-525, 1993.

Jensen, J. R.: Introductory digital image processing: a remote sensing perspective, 3rd ed., Prentice Hall series in geographic information science, Prentice Hall, Upper Saddle River, NY, 526 pp., 2005.

Ji, L., and Peters, A.: A spatial regression procedure for evaluating the relationship between AVHRR-NDVI and climate in the northern Great Plains, International Journal of Remote Sensing, 25, 297-311, 2004.

Jong, R., Schaepman, M. E., Furrer, R., Bruin, S., and Verburg, P. H.: Spatial relationship between climatologies and changes in global vegetation activity, Global change biology, 19, 1953-1964, 2013.

Jong, R., Verbesselt, J., Schaepman, M. E., and Bruin, S.: Trend changes in global greening and browning: contribution of short-term trends to longer-term change, Global Change Biology, 18, 642-655, 2012.

Joshi, C., and Mohanty, B. P.: Physical controls of near-surface soil moisture across varying spatial scales in an agricultural landscape during SMEX02, Water Resources Research, 46, 2010.

Kaiser, H. F.: The Application of Electronic Computers to Factor Analysis, Educational and Psychological Measurement, 20, 141-151, 1960.

Karlsen, S. R., Solheim, I., Beck, P. S., Høgda, K. A., Wielgolaski, F. E., and Tømmervik, H.: Variability of the start of the growing season in Fennoscandia, 1982–2002, International Journal of Biometeorology, 51, 513-524, 2007.

Kearney, J.: Food consumption trends and drivers, Philosophical transactions of the royal society B: biological sciences, 365, 2793-2807, 2010.

Keeling, R. F., Piper, S., and Heimann, M.: Global and hemispheric CO2 sinks deduced from changes in atmospheric O2 concentration, Nature, 381, 1996.

Kendall, M. G.: Rank correlation methods, 1948.

Kerr, Y. H., Waldteufel, P., Wigneron, J.-P., Delwart, S., Cabot, F., Boutin, J., Escorihuela, M.-J., Font, J., Reul, N., and Gruhier, C.: The smos mission: New tool for monitoring key elements of the global water cycle, Proceedings of the IEEE, 98, 666-687, 2010.

Khodaverdiloo, H., Homaee, M., van Genuchten, M. T., and Dashtaki, S. G.: Deriving and validating pedotransfer functions for some calcareous soils, Journal of Hydrology, 399, 93-99, 2011.

Kie, J. G., Bowyer, R. T., Nicholson, M. C., Boroski, B. B., and Loft, E. R.: Landscape heterogeneity at differing scales: effects on spatial distribution of mule deer, Ecology, 83, 530-544, 2002.

Kirschbaum, M. U.: The temperature dependence of soil organic matter decomposition, and the effect of global warming on soil organic C storage, Soil Biology and biochemistry, 27, 753-760, 1995.

Koenker, R.: A note on studentizing a test for heteroscedasticity, Journal of Econometrics, 17, 107-112, 1981.

Kramer, P.: Physiology of woody plants, Elsevier, 1979.

Kramer, P.: Water relations of plants, Academy Press, New York, 48, 1, 1983.

Kumar, S., Stohlgren, T. J., and Chong, G. W.: Spatial heterogeneity influences native and nonnative plant species richness, Ecology, 87, 3186-3199, 2006.

Lal, R., Hall, G., and Miller, F.: Soil degradation: I. Basic processes, Land Degradation & Development, 1, 51-69, 1989.

Lal, R.: Forest soils and carbon sequestration, Forest ecology and management, 220, 242-258, 2005.

Lambin, E., and Ehrlich, D.: The surface temperature-vegetation index space for land cover and land-cover change analysis, International journal of remote sensing, 17, 463-487, 1996.

Lambin, E. F.: Modelling and monitoring land-cover change processes in tropical regions, Progress in Physical Geography, 21, 375-393, 1997.

Lambin, E. F., Geist, H. J., and Lepers, E.: Dynamics of land-use and land-cover change in tropical regions Annual Review of Environment and Resources, 28, 205-241, 10.1146/annurev.energy.28.050302.105459, 2003.

Lambin, E. F., and Meyfroidt, P.: Land use transitions: Socio-ecological feedback versus socio-economic change, Land Use Policy, 27, 108-118, 10.1016/j.landusepol.2009.09.003, 2010.

Landin, M. G., and Bosart, L. F.: The diurnal variation of precipitation in California and Nevada, Monthly weather review, 117, 1801-1816, 1989.

Landis, J. R., and Koch, G. G.: The measurement of observer agreement for categorical data, biometrics, 159-174, 1977.

Lange, O. L., Lösch, R., Schulze, E.-D., and Kappen, L.: Responses of stomata to changes in humidity, Planta, 100, 76-86, 1971.

Larson, W. E., and Pierce, F. J.: The dynamics of soil quality as a measure of sustainable management. Pages 37-51, in: Defining soil quality for a sustainable environment, edited by: J.W. Doran, D. C. C., D.F. Bezdicek and B.A. Stewart, definingsoilqua, SSSA and ASA, Madison, WI, 1994.

Lavers, T.: 'Land grab' as development strategy? The political economy of agricultural investment in Ethiopia, Journal of Peasant Studies, 39, 105-132, 2012.

Law, B., Falge, E., Gu, L. v., Baldocchi, D., Bakwin, P., Berbigier, P., Davis, K., Dolman, A., Falk, M., and Fuentes, J.: Environmental controls over carbon dioxide and water vapor exchange of terrestrial vegetation, Agricultural and Forest Meteorology, 113, 97-120, 2002.

Leach, M., and Fairhead, J.: Challenging Neo-Malthusian Deforestation Analyses in West Africa's Dynamic Forest Landscapes, Population and Development Review, 26, 17-43, 2000.

Leenaars, J.: Africa soil profiles database, Version 1.1, A compilation of georeferenced and standardised legacy soil profile data for Sub-Saharan Africa (with dataset). ISRIC Report, 3, 2013.

Legates, D. R., and Willmott, C. J.: Mean seasonal and spatial variability in gauge-corrected, global precipitation, International Journal of Climatology, 10, 111-127, 1990.

Leh, M., Bajwa, S., and Chaubey, I.: Impact of land use change on erosion risk: an integrated remote sensing, geographic information system and modeling methodology, Land Degradation & Development, 24, 409-421. DOI: 410.1002/ldr.1137, 10.1002/ldr.1137, 2013.

Leij, F., Alves, W., Van Genuchten, M. T., and Williams, J.: The UNSODA Unsaturated Soil Hydraulic Database; User's Manual, Version 1.0, Rep. EPA/600/R-96, 95, 103, 1996.

Leij, F. J., Romano, N., Palladino, M., Schaap, M. G., and Coppola, A.: Topographical attributes to predict soil hydraulic properties along a hillslope transect, Water Resources Research, 40, 2004.

Leitao, A. B., and Ahern, J.: Applying landscape ecological concepts and metrics in sustainable landscape planning, Landscape and urban planning, 59, 65-93, 2002.

Leitão, A. B., Miller, J., Ahern, J., and McGarigal, K.: Measuring landscapes: A planner's handbook, Island press, 2012.

Lemenih, M., and Itanna, F.: Soil carbon stocks and turnovers in various vegetation types and arable lands along an elevation gradient in southern Ethiopia, Geoderma, 123, 177-188, 2004.

Lemenih, M., Kassa, H., Kassie, G., Abebaw, D., and Teka, W.: Resettlement and woodland management problems and options: A case study from northwestern Ethiopia, Land Degradation & Development, 2012.

Lemma, B.: Human intervention in two lakes: Lessons from Lakes Alemaya and Hora-Kilole, Proceedings of the National Consultative workshop on the Ramsar Convention and Ethiopia, 2004, 18-19,

Lepsch, I., Menk, J., and Oliveira, J. d.: Carbon storage and other properties of soils under agriculture and natural vegetation in Sao Paulo State, Brazil, Soil Use and Management, 10, 34-42, 1994.

Lesschen, J. P., Peter H. Verburg, and Staal, S. J.: Statistical methods for analysing the spatial dimension of changes in land use and farming systems-LUCC Report Series No. 7, LUCC Focus 3 Office and ILRI 2005, 80 pp., 2005.

Letts, M. G., Lafleur, P. M., and Roulet, N. T.: On the relationship between cloudiness and net ecosystem carbon dioxide exchange in a peatland ecosystem, Ecoscience, 12, 53-59, 2005.

Liao, K.-H., Xu, S.-H., Wu, J.-C., Ji, S.-H., and Lin, Q.: Assessing Soil Water Retention Characteristics and Their Spatial Variability Using Pedotransfer Functions, Pedosphere, 21, 413-422, 2011.

Lilly, A.: A description of the HYPRES database (Hydraulic Properties of European Soils), The use of pedotransfer functions in soil hydrology research. Proc. Worksh. of the Project "Using Existing Soil Data to Derive Hydraulic Parameters for Simulation Modelling in Environmental Studies and in Land Use Planning," 2nd, Orleans, France, 10-12, 1996.

Lin, D., Xia, J., and Wan, S.: Climate warming and biomass accumulation of terrestrial plants: a meta-analysis, New Phytologist, 188, 187-198, 2010.

Lin, H. S., McInnes, K. J., Wilding, L. P., and Hallmark, C. T.: Effects of Soil Morphology on Hydraulic Properties II. Hydraulic Pedotransfer Functions Contribution from the Texas Agric. Exp. Stn., The Texas A&M Univ. System, Soil Sci. Soc. Am. J., 63, 955-961, 10.2136/sssaj1999.634955x, 1999.

Linder, W.: Digital Photogrammetry: A Practical Course, Springer, 2009.

Liu, Y., Parinussa, R., Dorigo, W., De Jeu, R., Wagner, W., Van Dijk, A., McCabe, M., and Evans, J.: Developing an improved soil moisture dataset by blending passive and active microwave satellite-based retrievals, Hydrology and Earth System Sciences, 15, 425-436, 2011.

Liu, Y., Dorigo, W., Parinussa, R., De Jeu, R., Wagner, W., McCabe, M., Evans, J., and Van Dijk, A.: Trend-preserving blending of passive and active microwave soil moisture retrievals, Remote Sensing of Environment, 123, 280-297, 2012a.

Liu, Z., Yao, Z., Huang, H., Wu, S., and Liu, G.: Land use and climate changes and their impacts on runoff in the Varlung Zangbo river basin, China, Land Degradation & Development, DOI: 10.1002/ldr.1159, 10.1002/ldr.1159, 2012b.

Loiselle, S., Cózar, A., van Dam, A., Kansiime, F., Kelderman, P., Saunders, M., and Simonit, S.: Tools for wetland ecosystem resource management in East Africa: focus on the Lake Victoria papyrus wetlands, in: Wetlands and natural resource management, Springer, 97-121, 2006.

Long, S. P., Ainsworth, E. A., Rogers, A., and Ort, D. R.: Rising atmospheric carbon dioxide: plants FACE the Future, Annu. Rev. Plant Biol., 55, 591-628, 2004.

Lu, D., Mausel, P., Brondizio, E., and Moran, E.: Change detection techniques, International journal of remote sensing, 25, 2365-2401, 2004.

Lucht, W., Prentice, I. C., Myneni, R. B., Sitch, S., Friedlingstein, P., Cramer, W., Bousquet, P., Buermann, W., and Smith, B.: Climatic control of the high-latitude vegetation greening trend and Pinatubo effect, Science, 296, 1687-1689, 2002.

Luo, Y., Gerten, D., Le Maire, G., Parton, W. J., Weng, E., Zhou, X., Keough, C., Beier, C., Ciais, P., and Cramer, W.: Modeled interactive effects of precipitation, temperature, and [CO2] on ecosystem carbon and water dynamics in different climatic zones, Global Change Biology, 14, 1986-1999, 2008.

Mac, M. J., Opler, P. A., Haecker, C. E., and Doran, P. D.: Status and trends of the nation's biological resources, US Department of the Interior, US Geological Survey Reston (VA), 1998.

Madden, R. A., and Julian, P. R.: Detection of a 40-50 day oscillation in the zonal wind in the tropical Pacific, Journal of the Atmospheric Sciences, 28, 702-708, 1971.

Mahmood, R., Leeper, R., and Quintanar, A. I.: Sensitivity of planetary boundary layer atmosphere to historical and future changes of land use/land cover, vegetation fraction, and soil moisture in Western Kentucky, USA, Global and Planetary Change, 78, 36-53, 2011.

Malthus, T. R.: An Essay on the Principle of Population, Sixth Edition ed., History of economics series, John Murray, London, 1826.

Manandhar, R., Odeh, I. O. A., and Pontius Jr, R. G.: Analysis of twenty years of categorical land transitions in the Lower Hunter of New South Wales, Australia, Agriculture, Ecosystems & Environment, 135, 336-346, 2010.

Mann, H. B.: Nonparametric tests against trend, Econometrica: Journal of the Econometric Society, 245-259, 1945.

Mapa, R. B.: Effect of reforestation using Tectona grandis on infiltration and soil water retention, Forest Ecology and Management, 77, 119-125, 1995.

Martiny, N., Camberlin, P., Richard, Y., and Philippon, N.: Compared regimes of NDVI and rainfall in semi-arid regions of Africa, International Journal of Remote Sensing, 27, 5201-5223, 2006.

Matsoukas, C., Benas, N., Hatzianastassiou, N., Pavlakis, K., Kanakidou, M., and Vardavas, I.: Potential evaporation trends over land between 1983–2008: driven by radiative fluxes or vapour-pressure deficit?, Atmospheric Chemistry and Physics, 11, 7601-7616, 2011.

Matsushita, B., Xu, M., and Fukushima, T.: Characterizing the changes in landscape structure in the Lake Kasumigaura Basin, Japan using a high-quality GIS dataset, Landscape and urban planning, 78, 241-250, 2006.

Matthews, E., and Fung, I.: Methane emission from natural wetlands: global distribution, area, and environmental characteristics of sources, Global biogeochemical cycles, 1, 61-86, 1987.

McClain, M.: Balancing Water Resources Development and Environmental Sustainability in Africa: A Review of Recent Research Findings and Applications, AMBIO, 42, 549-565, 10.1007/s13280-012-0359-1, 2013.

McDonald, E. R., Wu, X., Caccetta, P., and Campbell, N.: Illumination correction of landsat TM data in South East NSW, Tenth Australasian Remote Sensing and Photgrammetrey Conference, 2000, 1-13,

McGarigal, K., Cushman, S. A., Neel, M. C. & Ene, E.: FRAGSTATS: Spatial Pattern Analysis Program for Categorical Maps. , Computer software program produced by the authors at the University of Massachusetts, Amherst., USA, 2002.

McHugh, O., McHugh, A., Eloundou-Enyegue, P., and Steenhuis, T.: Integrated qualitative assessment of wetland hydrological and land cover changes in a data scarce dry Ethiopian highland watershed, Land Degradation & Development, 18, 643-658, 2007.

McKenzie, N.: Australian soils and landscapes: an illustrated compendium, CSIRO publishing, Australia, 2004.

McKergow, L., Gallant, J., and Dowling, T.: Modelling wetland extent using terrain indices, Lake Taupo, NZ, Proceedings of MODSIM 2007 International Congress on Modelling and Simulation, Modelling and Simulation Society of Australia and New Zealand, 2007, 74-80,

McMahon, S. M., Parker, G. G., and Miller, D. R.: Evidence for a recent increase in forest growth, Proceedings of the National Academy of Sciences, 107, 3611-3615, 2010.

MEA, A. M. E.: Ecosystems and human well-being: desertification synthesis, World Resources Institute, 2005.

Menard, S.: Applied Logistic Regression Analysis, Quantitative Applications in the Social Sciences, SAGE Publications, 111 pp., 2001.

Menard, S.: Logistic Regression: From Introductory to Advanced Concepts and Applications, SAGE Publications, 2009.

Mengistu, D., Bewket, W., and Lal, R.: Recent spatiotemporal temperature and rainfall variability and trends over the Upper Blue Nile River Basin, Ethiopia, International Journal of Climatology, 34, 2278-2292, 10.1002/joc.3837, 2014.

Menzel, A., Sparks, T. H., Estrella, N., Koch, E., Aasa, A., Ahas, R., ALM-KÜBLER, K., Bissolli, P., Braslavská, O. g., and Briede, A.: European phenological response to climate change matches the warming pattern, Global change biology, 12, 1969-1976, 2006.

Merlin, O., Al Bitar, A., Walker, J. P., and Kerr, Y.: An improved algorithm for disaggregating microwave-derived soil moisture based on red, near-infrared and thermal-infrared data, Remote sensing of Environment, 114, 2305-2316, 2010.

Metternicht, G., Zinck, J., Blanco, P., and Del Valle, H.: Remote sensing of land degradation: Experiences from Latin America and the Caribbean, Journal of environmental quality, 39, 42-61, 2010.

Meyer, P., Itten, K. I., Kellenberger, T., Sandmeier, S., and Sandmeier, R.: Radiometric corrections of topographically induced effects on Landsat TM data in an alpine environment, ISPRS Journal of Photogrammetry and Remote Sensing, 48, 17-28, 1993.

Miller, A.: Subset Selection in Regression, Second Editon, 2002.

Milne, B., Gupta, V., and Restrepo, C.: scale invariant coupling of plants, water, energy, and terrain, Ecoscience, 2002.

Min, Q.: Impacts of aerosols and clouds on forest-atmosphere carbon exchange, Journal of Geophysical Research: Atmospheres (1984–2012), 110, 2005.

Minasny, B., McBratney, A. B., and Bristow, K. L.: Comparison of different approaches to the development of pedotransfer functions for water-retention curves, Geoderma, 93, 225-253, 1999.

Misra, R. V., Lesschen, J. P., Smaling, E., and Roy, R. N.: Assessment of soil nutrient balance: approaches and methodologies, FAO Fertilizer & Plant Nutrition Bulletin, FAO, Rome, Italy, 2003.

Mohamed, Y., Bastiaanssen, W., and Savenije, H.: Spatial variability of evaporation and moisture storage in the swamps of the upper Nile studied by remote sensing techniques, Journal of Hydrology, 289, 145-164, 2004.

Mohamed, Y., Van den Hurk, B., Savenije, H., and Bastiaanssen, W.: Impact of the Sudd wetland on the Nile hydroclimatology, Water resources research, 41, 2005.

Mohanty, B., Shouse, P., and van Genuchten, M. T.: Spatio-temporal dynamics of water and heat in a field soil, Soil and Tillage Research, 47, 133-143, 1998.

Mohanty, B., and Skaggs, T.: Spatio-temporal evolution and time-stable characteristics of soil moisture within remote sensing footprints with varying soil, slope, and vegetation, Advances in Water Resources, 24, 1051-1067, 2001.

Moran, M., Clarke, T., Inoue, Y., and Vidal, A.: Estimating crop water deficit using the relation between surface-air temperature and spectral vegetation index, Remote sensing of environment, 49, 246-263, 1994.

Moran, P. A.: Notes on continuous stochastic phenomena, Biometrika, 17-23, 1950.

Müller, R., Müller, D., Schierhorn, F., and Gerold, G.: Spatiotemporal modeling of the expansion of mechanized agriculture in the Bolivian lowland forests, Applied Geography, 31, 631-640, 10.1016/j.apgeog.2010.11.018, 2011.

Mulugeta, S., Abbot, P., Sishaw, T., and Hailu, A.: Socio-economic determinants of wetland use in the Metu and Yayu-Hurumu Weredas of Illubabor, Report of the Ethiopian Wetlands Research Programme: Sustainable wetland management in Illubabor Zone, South-west Ethiopia, 36, 2000.

Muñoz-Rojas, M., Jordán, A., Zavala, L. M., La Rosa, D. D., Abd-Elmabod, S. K., and Anaya-Romero, M.: Impact of land use and land cover changes on organic carbon stocks in mediterranean soils, Land Degradation & Development, DOI: 10.1002/ldr.2194, 10.1002/ldr.2194, 2012.

Murty, D., Kirschbaum, M. U., Mcmurtrie, R. E., and Mcgilvray, H.: Does conversion of forest to agricultural land change soil carbon and nitrogen? A review of the literature, Global Change Biology, 8, 105-123, 2002.

Myneni, R., Keeling, C., Tuckers, C., Asrar, G., and Nemani, R.: Increased plant growth in the northern high latitudes from 1981 to 1991, Nature, 386, 698-702, 1997.

Nabhan, H.: Soil fertility management in support of food security in sub-Saharan Africa, FAO, 2001.

Nagol, J. R., Vermote, E. F., and Prince, S. D.: Effects of atmospheric variation on AVHRR NDVI data, Remote Sensing of Environment, 113, 392-397, 2009.

Narumalani, S., Mishra, D. R., and Rothwell, R. G.: Change detection and landscape metrics for inferring anthropogenic processes in the greater EFMO area, Remote Sensing of Environment, 91, 478-489, 2004.

NCFSE, N. C. f. F. S. i. E.: Voluntary Resettlement Programme (Access to improved land), vol. II, Addis Ababa, Ethiopia, 2003.

Neel, M. C., McGarigal, K., and Cushman, S. A.: Behavior of class-level landscape metrics across gradients of class aggregation and area, Landscape ecology, 19, 435-455, 2004.

Neeti, N., and Eastman, J. R.: A contextual mann-kendall approach for the assessment of trend significance in image time series, Transactions in GIS, 15, 599-611, 2011.

Neill, C., Melillo, J. M., Steudler, P. A., Cerri, C. C., de Moraes, J. F., Piccolo, M. C., and Brito, M.: Soil carbon and nitrogen stocks following forest clearing for pasture in the southwestern Brazilian Amazon, Ecological Applications, 7, 1216-1225, 1997.

Nelson, D. W., and Sommers, L. E.: Total Carbon, Organic Carbon, and Organic Matter1, Methods of Soil Analysis. Part 2. Chemical and Microbiological Properties, agronomymonogra, 539-579, 10.2134/agronmonogr9.2.2ed.c29, 1982.

Nelson, J. A., Morgan, J. A., LeCain, D. R., Mosier, A. R., Milchunas, D. G., and Parton, B. A.: Elevated CO2 increases soil moisture and enhances plant water relations in a long-term field study in semi-arid shortgrass steppe of Colorado, Plant and Soil, 259, 169-179, 2004.

Nemani, R. R., Keeling, C. D., Hashimoto, H., Jolly, W. M., Piper, S. C., Tucker, C. J., Myneni, R. B., and Running, S. W.: Climate-driven increases in global terrestrial net primary production from 1982 to 1999, science, 300, 1560-1563, 2003.

Nemes, A., Rawls, W., Pachepsky, Y. A., and van Genuchten, M. T.: Sensitivity analysis of the nonparametric nearest neighbor technique to estimate soil water retention, Vadose Zone Journal, 5, 1222-1235, 2006.

Nemes, A., Rawls, W. J., and Pachepsky, Y. A.: Use of the nonparametric nearest neighbor approach to estimate soil hydraulic properties, Soil Science Society of America Journal, 70, 327-336, 2006.

Nemes, A., Roberts, R., Rawls, W., Pachepsky, Y. A., and van Genuchten, M. T.: Software to estimate- 33 and- 1500kPa soil water retention using the non-parametric k-Nearest Neighbor technique, Environmental Modelling & Software, 23, 254-255, 2008.

Nemes, A., Schaap, M., Leij, F., and Wösten, J.: Description of the unsaturated soil hydraulic database UNSODA version 2.0, Journal of Hydrology, 251, 151-162, 2001.

Nemes, A., Schaap, M., and Wösten, J.: Functional evaluation of pedotransfer functions derived from different scales of data collection, Soil Science Society of America Journal, 67, 1093-1102, 2003.

Nemes, A., Wösten, J., Bouma, J., and Várallyay, G.: Soil water balance scenario studies using predicted soil hydraulic parameters, Hydrological processes, 20, 1075-1094, 2006.

New, M., Hulme, M., and Jones, P.: Representing twentieth-century space-time climate variability. Part I: Development of a 1961-90 mean monthly terrestrial climatology, Journal of Climate, 12, 829-856, 1999.

Niemeijer, D.: Developing indicators for environmental policy: data-driven and theory-driven approaches examined by example, Environmental Science & Policy, 5, 91-103, 2002.

Niu, S., Wu, M., Han, Y., Xia, J., Li, L., and Wan, S.: Water-mediated responses of ecosystem carbon fluxes to climatic change in a temperate steppe, New Phytologist, 177, 209-219, 2008.

Niyogi, D., Mahmood, R., and Adegoke, J.: Land-Use/Land-Cover Change and Its Impacts on Weather and Climate, Boundary-Layer Meteorology, 133, 297-298, 10.1007/s10546-009-9437-8, 2009.

Njoku, E. G., Jackson, T. J., Lakshmi, V., Chan, T. K., and Nghiem, S. V.: Soil moisture retrieval from AMSR-E, Geoscience and Remote Sensing, IEEE Transactions on, 41, 215-229, 2003.

Norby, R. J., and Luo, Y.: Evaluating ecosystem responses to rising atmospheric CO2 and global warming in a multi-factor world, New Phytologist, 162, 281-293, 2004.

Nyssen, J., Poesen, J., Moeyersons, J., Deckers, J., Haile, M., and Lang, A.: Human impact on the environment in the Ethiopian and Eritrean highlands—a state of the art, Earth-Science Reviews, 64, 273-320, 2004.

Obalum, S. E., and Obi, M. E.: Moisture characteristics and their point pedotransfer functions for coarse-textured tropical soils differing in structural degradation status, Hydrological Processes, 27, 2721-2735, 10.1002/hyp.9398, 2013.

OECD.: Guidelines for aid agencies for improved conservation and sustainable use of tropical and sub-tropical wetlands, OECD, 1996.

Olsson, H.: Regression functions for multitemporal relative calibration of Thematic Mapper data over boreal forest, Remote Sensing of Environment, 46, 89-102, 1993.

Olsson, L., Eklundh, L., and Ardö, J.: A recent greening of the Sahel—trends, patterns and potential causes, Journal of Arid Environments, 63, 556-566, 2005.

Olthof, I., Pouliot, D., Fernandes, R., and Latifovic, R.: Landsat-7 ETM+ radiometric normalization comparison for northern mapping applications, Remote Sensing of Environment, 95, 388-398, 2005.

Onema, J.-M. K., and Taigbenu, A.: NDVI–rainfall relationship in the Semliki watershed of the equatorial Nile, Physics and Chemistry of the Earth, Parts A/B/C, 34, 711-721, 2009.

Ordoyne, C., and Friedl, M. A.: Using MODIS data to characterize seasonal inundation patterns in the Florida Everglades, Remote Sensing of Environment, 112, 4107-4119, 2008.

Ozesmi, S. L., and Bauer, M. E.: Satellite remote sensing of wetlands, Wetlands ecology and management, 10, 381-402, 2002.

Pachepsky, Y., and Schaap, M.: Data mining and exploration techniques, Developments in Soil Science, 30, 21-32, 2004.

Pachepsky, Y. A., and Rawls, W. J.: Accuracy and Reliability of Pedotransfer Functions as Affected by Grouping Soils 1 Currently classified as Haplustepts, Soil Sci. Soc. Am. J., 63, 1748-1757, 1999.

Pachepsky, Y. A., and Rawls, W.: Soil structure and pedotransfer functions, European Journal of Soil Science, 54, 443-452, 2003.

Palumbo, A. D., Vitale, D., Campi, P., and Mastrorilli, M.: Time trend in reference evapotranspiration: analysis of a long series of agrometeorological measurements in Southern Italy, Irrigation and Drainage Systems, 25, 395-411, 2011.

Pantaleoni, E., Wynne, R., Galbraith, J., and Campbell, J.: Mapping wetlands using ASTER data: a comparison between classification trees and logistic regression, International Journal of Remote Sensing, 30, 3423-3440, 2009.

Paolini, L., Grings, F., Sobrino, J. A., Jiménez Muñoz, J. C., and Karszenbaum, H.: Radiometric correction effects in Landsat multi-date/multi-sensor change detection studies, International Journal of Remote Sensing, 27, 685-704, 2006.

Parry, M. L.: Climate Change 2007: impacts, adaptation and vulnerability: contribution of Working Group II to the fourth assessment report of the Intergovernmental Panel on Climate Change, Cambridge University Press, 2007.

Pender, J., Berhanu Gebremedhin, Samuel Benin, and Ehui, S.: Strategies for Sustainable Agricultural Development in the Ethiopian Highlands, American Journal of Agricultural Economics, 83, 5, 1231-1240, 2001.

Peng, S., Chen, A., Xu, L., Cao, C., Fang, J., Myneni, R. B., Pinzon, J. E., Tucker, C. J., and Piao, S.: Recent change of vegetation growth trend in China, Environmental Research Letters, 6, 044027, 2011.

Peterson, T. C., Golubev, V. S., and Groisman, P. Y.: Evaporation losing its strength, Nature, 377, 687-688, 1995.

Pettitt, A.: A non-parametric approach to the change-point problem, Applied statistics, 126-135, 1979.

Philip, J. R.: The theory of infiltration: 4. Sorptivity and algebraic infiltration equations, Soil science, 84, 257-264, 1957.

Pongratz, J., Reick, C. H., Raddatz, T., and Claussen, M.: Effects of anthropogenic land cover change on the carbon cycle of the last millennium, Global Biogeochemical Cycles, 23, 1-13, 10.1029/2009gb003488, 2009.

Pontius Jr, R. G., and Schneider, L. C.: Land-cover change model validation by an ROC method for the Ipswich watershed, Massachusetts, USA, Agriculture, Ecosystems & Environment, 85, 239-248, 2001.

Pontius, R.: Detecting important categorical land changes while accounting for persistence, Agriculture, Ecosystems & Environment, 101, 251-268, 10.1016/j.agee.2003.09.008, 2004.

Price, J. C.: Using spatial context in satellite data to infer regional scale evapotranspiration, Geoscience and Remote Sensing, IEEE Transactions on, 28, 940-948, 1990.

Prince, S.: Satellite remote sensing of primary production: comparison of results for Sahelian grasslands 1981-1988, International Journal of Remote Sensing, 12, 1301-1311, 1991.

Prince, S., Becker-Reshef, I., and Rishmawi, K.: Detection and mapping of long-term land degradation using local net production scaling: Application to Zimbabwe, Remote Sensing of Environment, 113, 1046-1057, 2009.

Prince, S. D., and Goward, S. N.: Global primary production: a remote sensing approach, Journal of biogeography, 815-835, 1995.

Prince, S. D., Wessels, K. J., Tucker, C. J., and Nicholson, S. E.: Desertification in the Sahel: a reinterpretation of a reinterpretation, Global Change Biology, 13, 1308-1313, 2007.

Puyravaud, J.-P.: Standardizing the calculation of the annual rate of deforestation, Forest Ecology and Management, 177, 593-596, 10.1016/s0378-1127(02)00335-3, 2003.

Rahmato, D.: Searching for Tenure Security? The Land System and New Policy Initiatives in Ethiopia, Discussion Paper No. 12, Forum for Social Studies, Addis Abeba, Ethiopia, 2004.

Ramankutty, N., Gibbs, H. K., Achard, F., Defries, R., Foley, J. A., and Houghton, R.: Challenges to estimating carbon emissions from tropical deforestation, Global Change Biology, 13, 51-66, 2007.

Ramsar Convenon Secretariat, R. C. S.: The Ramsar convention manual: a guide to the convention on wetlands (Ramsar, Iran, 1971), Ramsar Convention Secretariat, Gland, Switzerland, 2006,

Raveh, E., Cohen, S., Raz, T., Yakir, D., Grava, A., and Goldschmidt, E.: Increased growth of young citrus trees under reduced radiation load in a semi-arid climate1, Journal of experimental botany, 54, 365-373, 2003.

Rawls, W., and Brakensiek, D.: Estimation of soil water retention and hydraulic properties, in: Unsaturated flow in hydrologic modeling, Springer, 275-300, 1989.

Rawls, W., Brakensiek, D., and Savabi, M.: Infiltration parameters for rangeland soils, Journal of Range Management, 139-142, 1989.

Rawls, W. J., Brakensiek, D. L., and Saxtonn, K. E.: Estimation of Soil Water Properties, Transactions of the ASAE 25, 1316-1320, 1982.

Rawls, W. J., Brakensiek, D. L., and Miller, N.: Green-Ampt infiltration parameters from soils data, Journal of hydraulic engineering, 109, 62-70, 1983.

Ray, R. L., Jacobs, J. M., and Cosh, M. H.: Landslide susceptibility mapping using downscaled AMSR-E soil moisture: A case study from Cleveland Corral, California, US, Remote sensing of environment, 114, 2624-2636, 2010.

Rayment, G. E., and Higginson, F. R.: Australian Laboratory Handbook of Soil and Water Chemical Methods, Inkata Press, Melbourne, 1992.

Reed, B. C., Brown, J. F., VanderZee, D., Loveland, T. R., Merchant, J. W., and Ohlen, D. O.: Measuring phenological variability from satellite imagery, Journal of Vegetation Science, 5, 703-714, 1994.

Reimold, R. J.: Wetlands functions and values, Applied Wetlands Science and Technology. Lewis: Boca Raton, 55-78, 1994.

Richardson, A. D., Black, T. A., Ciais, P., Delbart, N., Friedl, M. A., Gobron, N., Hollinger, D. Y., Kutsch, W. L., Longdoz, B., and Luyssaert, S.: Influence of spring and autumn phenological transitions on forest ecosystem productivity, Philosophical Transactions of the Royal Society B: Biological Sciences, 365, 3227-3246, 2010.

Richter, G., and Negendank, J. F.: Soil erosion processes and their measurement in the German area of the Moselle river, Earth Surface Processes, 2, 261-278, 1977.

Riitters, K. H., O'neill, R., Hunsaker, C., Wickham, J. D., Yankee, D., Timmins, S., Jones, K., and Jackson, B.: A factor analysis of landscape pattern and structure metrics, Landscape ecology, 10, 23-39, 1995.

Riley, G. T., Landin, M. G., and Bosart, L. F.: The diurnal variability of precipitation across the central Rockies and adjacent Great Plains, Monthly weather review, 115, 1161-1172, 1987.

Rocha, A. V., Su, H.-B., Vogel, C. S., Schmid, H. P., and Curtis, P. S.: Photosynthetic and water use efficiency responses to diffuse radiation by an aspen-dominated northern hardwood forest, Forest Science, 50, 793-801, 2004.

Roderick, M. L., and Farquhar, G. D.: Changes in Australian pan evaporation from 1970 to 2002, International Journal of Climatology, 24, 1077-1090, 2004.

Roerink, G., Menenti, M., and Verhoef, W.: Reconstructing cloudfree NDVI composites using Fourier analysis of time series, International Journal of Remote Sensing, 21, 1911-1917, 2000.

Ronald Eastman, J., Sangermano, F., Ghimire, B., Zhu, H., Chen, H., Neeti, N., Cai, Y., Machado, E. A., and Crema, S. C.: Seasonal trend analysis of image time series, International Journal of Remote Sensing, 30, 2721-2726, 2009.

Running, S. W., Nemani, R. R., Heinsch, F. A., Zhao, M., Reeves, M., and Hashimoto, H.: A continuous satellite-derived measure of global terrestrial primary production, Bioscience, 54, 547-560, 2004.

Russell, G. C., and Green, K.: Assessing the accuracy of remotely sensed data: principles and practices, Mapping science series, Lewis Publications, 1999.

Rustad, L., Campbell, J., Marion, G., Norby, R., Mitchell, M., Hartley, A., Cornelissen, J., and Gurevitch, J.: A meta-analysis of the response of soil respiration, net nitrogen mineralization, and aboveground plant growth to experimental ecosystem warming, Oecologia, 126, 543-562, 2001.

Salinger, M. J.: Climate variability and change: past, present and future–an overview, Climatic Change, 70, 9-29, 2005.

Sandholt, I., Rasmussen, K., and Andersen, J.: A simple interpretation of the surface temperature/vegetation index space for assessment of surface moisture status, Remote Sensing of environment, 79, 213-224, 2002.

Savenije, H. H.: New definitions for moisture recycling and the relationship with land-use changes in the Sahel, Journal of Hydrology, 167, 57-78, 1995.

Savenije, H. H., and Van Der Zaag, P.: Water as an economic good and demand management paradigms with pitfalls, Water international, 27, 98-104, 2002.

Savva, Y., Szlavecz, K., Carlson, D., Gupchup, J., Szalay, A., and Terzis, A.: Spatial patterns of soil moisture under forest and grass land cover in a suburban area, in Maryland, USA, Geoderma, 192, 202-210, 2013.

Scanlon, T. M., and Albertson, J. D.: Canopy scale measurements of CO_2 and water vapor exchange along a precipitation gradient in southern Africa, Global Change Biology, 10, 329-341, 2004.

Schaap, M., Leij, F., and Van Genuchten, M. T.: A bootstrap-neural network approach to predict soil hydraulic parameters, Characterization and Measurements of the Hydraulic Properties of Unsaturated Porous Media, 1237-1250, 1999.

Schaap, M. G., and Leij, F. J.: Using neural networks to predict soil water retention and soil hydraulic conductivity, Soil and Tillage Research, 47, 37-42, 1998a.

Schaap, M. G., and Leij, F. J.: Database-related accuracy and uncertainty of pedotransfer functions, Soil Science, 163, 765-779, 1998b.

Schaap, M. G.: Accuracy and uncertainty in PTF predictions, Developments in soil science, 30, 33-43, 2004.

Schaap, M. G., and Bouten, W.: Modeling water retention curves of sandy soils using neural networks, Water Resources Research, 32, 3033-3040, 1996.

Schaap, M. G., and Leij, F. J.: Improved prediction of unsaturated hydraulic conductivity with the Mualem-van Genuchten model, Soil Science Society of America Journal, 64, 843-851, 2000.

Schaap, M. G., Leij, F. J., and van Genuchten, M. T.: Neural network analysis for hierarchical prediction of soil hydraulic properties, Soil Science Society of America Journal, 62, 847-855, 1998.

Schaap, M. G., Leij, F. J., and van Genuchten, M. T.: rosetta: a computer program for estimating soil hydraulic parameters with hierarchical pedotransfer functions, Journal of Hydrology, 251, 163-176, 2001.

Schaap, M. G., and Van Genuchten, M. T.: A modified Mualem–van Genuchten formulation for improved description of the hydraulic conductivity near saturation, Vadose Zone Journal, 5, 27-34, 2006.

Schimel, D. S., Braswell, B., and Parton, W.: Equilibration of the terrestrial water, nitrogen, and carbon cycles, Proceedings of the National Academy of Sciences, 94, 8280-8283, 1997.

Schindler, S., Poirazidis, K., and Wrbka, T.: Towards a core set of landscape metrics for biodiversity assessments: A case study from Dadia National Park, Greece, Ecological indicators, 8, 502-514, 2008.

Schmullius, C., and Furrer, R.: Some critical remarks on the use of C-band radar data for soil moisture detection, International Journal of Remote Sensing, 13, 3387-3390, 1992.

Schneider, L. C., and Pontius Jr, R. G.: Modeling land-use change in the Ipswich watershed, Massachusetts, USA, Agriculture, Ecosystems & Environment, 85, 83-94, 2001.

Schroeder, T. A., Cohen, W. B., Song, C., Canty, M. J., and Yang, Z.: Radiometric correction of multi-temporal Landsat data for characterization of early successional forest patterns in western Oregon, Remote Sensing of Environment, 103, 16-26, 10.1016/j.rse.2006.03.008, 2006.

Schulze, E.-D., and Hall, A.: Stomatal responses, water loss and CO2 assimilation rates of plants in contrasting environments, in: Physiological plant ecology II, Springer, 181-230, 1982.

Schuur, E. A.: Productivity and global climate revisited: the sensitivity of tropical forest growth to precipitation, Ecology, 84, 1165-1170, 2003.

Schwinning, S., and Sala, O. E.: Hierarchy of responses to resource pulses in arid and semi-arid ecosystems, Oecologia, 141, 211-220, 2004.

Sen, P. K.: Estimates of the regression coefficient based on Kendall's tau, Journal of the American Statistical Association, 63, 1379-1389, 1968.

Seneviratne, S. I., Corti, T., Davin, E. L., Hirschi, M., Jaeger, E. B., Lehner, I., Orlowsky, B., and Teuling, A. J.: Investigating soil moisture–climate interactions in a changing climate: A review, Earth-Science Reviews, 99, 125-161, 2010.

Serneels, S., and Lambin, E. F.: Proximate causes of land-use change in Narok District, Kenya: a spatial statistical model, Agriculture, Ecosystems & Environment, 85, 65-81, 10.1016/s0167-8809(01)00188-8, 2001.

Sheather, S.: A modern Approach to Regression with R, Springer, 2009.

Sherry, R. A., Weng, E., ARNONE III, J. A., Johnson, D. W., Schimel, D. S., Verburg, P. S., Wallace, L. L., and Luo, Y.: Lagged effects of experimental warming and doubled precipitation on annual and seasonal aboveground biomass production in a tallgrass prairie, Global Change Biology, 14, 2923-2936, 2008.

Shirley, S. M., and Smith, J. N.: Bird community structure across riparian buffer strips of varying width in a coastal temperate forest, Biological Conservation, 125, 475-489, 2005.

Shukla, M., Lal, R., and Ebinger, M.: Determining soil quality indicators by factor analysis, Soil and Tillage Research, 87, 194-204, 2006.

Shukla, M. K.: Introduction to Soil Hydrology: Processes and Variability of Hydrological Properties, in: Soil Hydrology, Land Use and Agriculture: Measurement and Modelling, edited by: Shukla, M. K., CABI, USA, 2011.

Shuttleworth, W. J., and Maidment, D.: Evaporation, McGraw-Hill Inc., 1992.

Sikka, A., Samra, J., Sharda, V., Samraj, P., and Lakshmanan, V.: Low flow and high flow responses to converting natural grassland into blue gum (Eucalyptus globulus) in Nilgiris watersheds of South India, Journal of Hydrology, 270, 12-26, 2003.

Singh, A.: Review article digital change detection techniques using remotely-sensed data, International journal of remote sensing, 10, 989-1003, 1989.

Slayback, D. A., Pinzon, J. E., Los, S. O., and Tucker, C. J.: Northern hemisphere photosynthetic trends 1982–99, Global Change Biology, 9, 1-15, 2003.

Smith, J., Lin, T. L., and Ranson, K.: The Lambertian assumption and Landsat data, Photogrammetric Engineering and Remote Sensing, 46, 1183-1189, 1980.

Song, C., Woodcock, C. E., Seto, K. C., Lenney, M. P., and Macomber, S. A.: Classification and change detection using Landsat TM data: when and how to correct atmospheric effects?, Remote sensing of Environment, 75, 230-244, 2001.

Sonneveld, B. G. J. S.: Land under pressure: the impact of water erosion on food production in Ethiopia, Shaker, 2002.

Southworth, J., Munroe, D., and Nagendra, H.: Land cover change and landscape fragmentation—comparing the utility of continuous and discrete analyses for a western Honduras region, Agriculture, ecosystems & environment, 101, 185-205, 2004.

Sparks, T. H., Menzel, A., and Stenseth, N. C.: European cooperation in plant phenology, Clim Res, 39, 175-177, 2009.

Starr, G.: Assessing temporal stability and spatial variability of soil water patterns with implications for precision water management, Agricultural Water Management, 72, 223-243, 2005.

Stehman, S. V.: Selecting and interpreting measures of thematic classification accuracy, Remote Sensing of Environment, 62, 77-89, 10.1016/s0034-4257(97)00083-7, 1997.

Stephens, M. A.: EDF statistics for goodness of fit and some comparisons, Journal of the American statistical Association, 69, 730-737, 1974.

Sterling, S. M., Ducharne, A., and Polcher, J.: The impact of global land-cover change on the terrestrial water cycle, Nature Climate Change, 3, 385-390, 2013.

Stevens, J.: Applied multivariate statistics for the social sciences (5th ed.), Routledge/Taylor & Francis Group, New York, US, 2009.

Tabari, H., and Talaee, P. H.: Local calibration of the Hargreaves and Priestley-Taylor equations for estimating reference evapotranspiration in arid and cold climates of Iran based on the Penman-Monteith model, Journal of Hydrologic Engineering, 16, 837-845, 2011.

Tabari, H., Aeini, A., Talaee, P. H., and Some'e, B. S.: Spatial distribution and temporal variation of reference evapotranspiration in arid and semi-arid regions of Iran, Hydrological Processes, 26, 500-512, 2012.

Tabari, H., and Aghajanloo, M. B.: Temporal pattern of aridity index in Iran with considering precipitation and evapotranspiration trends, International Journal of Climatology, 33, 396-409, 2013.

Taboada, M. A., and Lavado, R. S.: Grazing effects of the bulk density in a Natraquoll of the flooding Pampa of Argentina, Journal of Range Management, 500-503, 1988.

Tamari, S., Wösten, J., and Ruiz-Suarez, J.: Testing an artificial neural network for predicting soil hydraulic conductivity, Soil Science Society of America Journal, 60, 1732-1741, 1996.

Tanner, C. B., and Sinclair, T. R.: Efficient Water Use in Crop Production: Research or Re-Search?1, in: Limitations to Efficient Water Use in Crop Production, edited by: Taylor, H. M., Wayne, J. R., and Thomas, S. R., American Society of Agronomy, Crop Science Society of America, Soil Science Society of America, 1-27, 1983.

Tansey, K., and Millington, A.: Investigating the potential for soil moisture and surface roughness monitoring in drylands using ERS SAR data, International Journal of remote sensing, 22, 2129-2149, 2001.

TECSULT: Ethiopian Energy II Project. Woody Biomass Inventory and Strategic Planning Project (WBISPP) - Phase 2 - Terminal Report, Ministry of Agriculture, Addis Ababa, Ethiopia, 2004.

Tefera, M., Chernet, T., and Haro, W.: Explanation of the Geological Map of Ethiopia: Scale 1: 2,000,000, EIGS Technical Publications Team, 1999.

Teferi, E.: The changing bio-physical environments of Mt.Choke Building climate resilient green economy in Mt. Choke, Debre Markos, 2011, 43 & 44, 2011.

Teferi, E., Bewket, W., Simane, B., Uhlenbrook, S., and Wenninger, J.: Effects of land use and land cover on selected soil quality indictors in the headwater area of the Blue Nile Basin of Ethiopia, Environmental Monitoring and Assessment (under review), 2015.

Teferi, E., Bewket, W., Uhlenbrook, S., and Wenninger, J.: Understanding recent land use and land cover dynamics in the source region of the Upper Blue Nile, Ethiopia: Spatially explicit statistical modeling of systematic transitions, Agriculture, Ecosystems & Environment, 165, 98-117, 2013.

Teferi, E., Uhlenbrook, S., and Bewket, W.: Inter-annual and seasonal trends of vegetation condition in the Upper Blue Nile (Abbay) basin: dual scale time series analysis, Earth System Dynamics Discussions, 6, 169-216, 2015.

Teferi, E., Uhlenbrook, S., Bewket, W., Wenninger, J., and Simane, B.: The use of remote sensing to quantify wetland loss in the Choke Mountain range, Upper Blue Nile basin, Ethiopia, Hydrol. Earth Syst. Sci., 14, 2415-2428, 10.5194/hess-14-2415-2010, 2010.

Teferi, E., Wenninger, J., and Uhlenbrook, S.: Predicting soil water retention characteristics in high altitude tropical soils – A case study from the Upper Blue Nile/Abbay, Ethiopia. , CATENA (under review), 2015a.

Teferi, E., Wenninger, J., and Uhlenbrook, S.: Monitoring soil moisture in response to land use and land cover changes using Remote Sensing and in-situ observations, International Journal of Applied Earth Observation and GeoInformation (under review), 2015b.

Tegene, B.: Land-Cover/Land-Use Changes in the Derekolli Catchment of the South Welo Zone of Amhara Region, Ethiopia, Eastern Africa Social Science Research Review, 18, 1-20, 10.1353/eas.2002.0005, 2002.

Teillet, P., Barker, J., Markham, B., Irish, R., Fedosejevs, G., and Storey, J.: Radiometric cross-calibration of the Landsat-7 ETM+ and Landsat-5 TM sensors based on tandem data sets, Remote sensing of Environment, 78, 39-54, 2001.

Teillet, P. M., Guindon, B., and Goodenough, D. G.: On the slope-aspect correction of multispectral scanner data, Canadian Journal of Remote Sensing, 8, 84-106, 1982.

Tekleab, S., Mohamed, Y., and Uhlenbrook, S.: Hydro-climatic trends in the Abay/Upper Blue Nile basin, Ethiopia, Physics and Chemistry of the Earth, Parts A/B/C, 61, 32-42, 2013.

Tekleab, S., Mohamed, Y., Uhlenbrook, S., and Wenninger, J.: Hydrologic responses to land cover change: the case of Jedeb meso-scale catchment, Abay/Upper Blue Nile basin, Ethiopia, Hydrological Processes, *28*(20), 5149-516, 2014.

Tekleab, S., Uhlenbrook, S., Mohamed, Y., Savenije, H. H. G., Temesgen, M., and Wenninger, J.: Water balance modeling of Upper Blue Nile catchments using a top-down approach, Hydrol. Earth Syst. Sci., 15, 2179-2193, 10.5194/hess-15-2179-2011, 2011.

Temesgen, M., Hoogmoed, W., Rockstrom, J., and Savenije, H.: Conservation tillage implements and systems for smallholder farmers in semi-arid Ethiopia, Soil and Tillage Research, 104, 185-191, 2009.

Tesfahunegn, G. B.: Soil Quality Indicators Response to Land Use and Soil Management Systems in Northern Ethiopia's Catchment, Land Degradation & Development, doi: 10.1002/ldr.2245, 2013.

Thenkabail, P. S., Schull, M., and Turral, H.: Ganges and Indus river basin land use/land cover (LULC) and irrigated area mapping using continuous streams of MODIS data, Remote Sensing of Environment, 95, 317-341, 2005.

Thiel, H.: A rank-invariant method of linear and polynomial regression analysis, Part 3, Proceedings of Koninalijke Nederlandse Akademie van Weinenschatpen A, 1950, 1397-1412,

Thenkabail, P. S., Schull, M., and Turral, H.: Ganges and Indus river basin land use/land cover (LULC) and irrigated area mapping using continuous streams of MODIS data, Remote Sensing of Environment, 95, 317-341, 2005.

Thomas, G. W.: Exchangeable cations, Methods of soil analysis. Part 2. Chemical and microbiological properties, 159-165, 1982.

Tian, H., Chen, G., Liu, M., Zhang, C., Sun, G., Lu, C., Xu, X., Ren, W., Pan, S., and Chappelka, A.: Model estimates of net primary productivity, evapotranspiration, and water use efficiency in the terrestrial ecosystems of the southern United States during 1895–2007, Forest ecology and management, 259, 1311-1327, 2010.

Tietje, O., and Tapkenhinrichs, M.: Evaluation of Pedo-Transfer Functions, Soil Sci. Soc. Am. J., 57, 1088-1095, 10.2136/sssaj1993.03615995005700040035x, 1993.

Tiner, R. W.: Estimated extent of geographically isolated wetlands in selected areas of the United States, Wetlands, 23, 636-652, 2003.

Tobler, W.: Non-isotropic geographic modeling, National Center for Geographic Information and Analysis, Santa BarbaraTechnical Report No. 93-1, 1993.

Tokola, T., Löfman, S., and Erkkilä, A.: Relative Calibration of Multitemporal Landsat Data for Forest Cover Change Detection, Remote Sensing of Environment, 68, 1-11, 10.1016/s0034-4257(98)00096-0, 1999.

Tomasella, J., Pachepsky, Y., Crestana, S., and Rawls, W.: Comparison of two techniques to develop pedotransfer functions for water retention, Soil Science Society of America Journal, 67, 1085-1092, 2003.

Tomasella, J., and Hodnett, M.: Pedotransfer functions for tropical soils, in: Developments in Soil Science, edited by: Pachepsky, Y., and Rawls, W. J., Elsevier, 415-429, 2004.

Tombul, M., Akyürek, Z., and Ünal Sorman, A.: Research Note: Determination of soil hydraulic properties using pedotransfer functions in a semi-arid basin, Turkey, Hydrol. Earth Syst. Sci., 8, 1200-1209, 10.5194/hess-8-1200-2004, 2004.

Tomer, M. D., and Schilling, K. E.: A simple approach to distinguish land-use and climate-change effects on watershed hydrology, Journal of Hydrology, 376, 24-33, 2009.

Toothaker, L. E.: Multiple comparisons for researchers, Sage Publications, Inc, 1991.

Torras, O., Gil-Tena, A., and Saura, S.: How does forest landscape structure explain tree species richness in a Mediterranean context?, Biodiversity and Conservation, 17, 1227-1240, 2008.

Townshend, J. R., and Justice, C. O.: Selecting the spatial resolution of satellite sensors required for global monitoring of land transformations, International Journal of Remote Sensing, 9, 187-236, 1988.

Töyrä, J., and Pietroniro, A.: Towards operational monitoring of a northern wetland using geomatics-based techniques, Remote Sensing of Environment, 97, 174-191, 2005.

Tsegaye, D., Moe, S. R., Vedeld, P., and Aynekulu, E.: Land-use/cover dynamics in Northern Afar rangelands, Ethiopia, Agriculture, Ecosystems & Environment, 139, 174-180, 10.1016/j.agee.2010.07.017, 2010.

Tucker, C., and Sellers, P.: Satellite remote sensing of primary production, International journal of remote sensing, 7, 1395-1416, 1986.

Tucker, C. J., Dregne, H. E., and Newcomb, W. W.: Expansion and Contraction of the Sahara Desert from 1980 to 1990, Science, 253, 299-300, 1991.

Tucker, C. J., Holben, B. N., Elgin Jr, J. H., and McMurtrey III, J. E.: Remote sensing of total dry-matter accumulation in winter wheat, Remote Sensing of Environment, 11, 171-189, 1981.

Tucker, C. J., Slayback, D. A., Pinzon, J. E., Los, S. O., Myneni, R. B., and Taylor, M. G.: Higher northern latitude normalized difference vegetation index and growing season trends from 1982 to 1999, International journal of biometeorology, 45, 184-190, 2001.

Tucker, C. J., Pinzon, J. E., Brown, M. E., Slayback, D. A., Pak, E. W., Mahoney, R., Vermote, E. F., and El Saleous, N.: An extended AVHRR 8-km NDVI dataset compatible with MODIS and SPOT vegetation NDVI data, International Journal of Remote Sensing, 26, 4485-4498, 2005.

Tukey, D. C. H. F. M. J. W.: Understanding robust and exploratory data analysis, Wiley, New York, 2000.

Turner , B. L., Skole, D. L., Sanderson, S., Fischer, G., Fresco, L. O., and Leemans, R.: Land-Use and Land-Cover Change: Science/Research Plan, IGBP, 1995.

Turner, B. L., Lambin, E. F., and Reenberg, A.: The emergence of land change science for global environmental change and sustainability, Proceedings of the National Academy of Sciences of the United States of America, 104, 20666-20671, 10.1073/pnas.0704119104, 2007.

Turner, M. G.: Landscape ecology: the effect of pattern on process, Annual review of ecology and systematics, 171-197, 1989.

Turner, M. G.: Landscape ecology: what is the state of the science?, Annual review of ecology, evolution, and systematics, 319-344, 2005.

Twarakavi, N. K., Šimůnek, J., and Schaap, M.: Development of pedotransfer functions for estimation of soil hydraulic parameters using support vector machines, Soil Science Society of America Journal, 73, 1443-1452, 2009.

Uhlenbrook, S.: Biofuel and water cycle dynamics: what are the related challenges for hydrological processes research?, Hydrological Processes, 21, 3647-3650, 2007.

Uhlenbrook, S., Mohamed, Y., and Gragne, A. S.: Analyzing catchment behavior through catchment modeling in the Gilgel Abay, Upper Blue Nile River Basin, Ethiopia, Hydrol. Earth Syst. Sci., 14, 2153-2165, 10.5194/hess-14-2153-2010, 2010.

Ulaby, F. T., Dubois, P. C., and Van Zyl, J.: Radar mapping of surface soil moisture, Journal of Hydrology, 184, 57-84, 1996.

UNEP: Global Environmental Outlook, New York, October, 2007.

Urban, O., JANOUŠ, D., Acosta, M., CZERNÝ, R., Markova, I., NavrATil, M., Pavelka, M., POKORNÝ, R., ŠPRTOVÁ, M., and Zhang, R.: Ecophysiological controls over the net ecosystem exchange of mountain spruce stand. Comparison of the response in direct vs. diffuse solar radiation, Global Change Biology, 13, 157-168, 2007.

USBR: Land and Water Resources of the Blue Nile Basin. , edited by: Reclamation, U. S. D. o. I. B. o., Washington, DC, USA, 1964.

Van Dam, A., Dardona, A., Kelderman, P., and Kansiime, F.: A simulation model for nitrogen retention in a papyrus wetland near Lake Victoria, Uganda (East Africa), Wetlands ecology and management, 15, 469-480, 2007.

van den Berg, M., Klamt, E., van Reeuwijk, L. P., and Sombroek, W. G.: Pedotransfer functions for the estimation of moisture retention characteristics of Ferralsols and related soils, Geoderma, 78, 161-180, 1997.

van der Zaag, P.: Viewpoint–Water Variability, Soil Nutrient Heterogeneity and Market Volatility–Why Sub-Saharan Africa's Green Revolution Will Be Location-Specific and Knowledge-Intensive, Water Alternatives, 3, 154-160, 2009.

van Genuchten, M. T.: A Closed-form Equation for Predicting the Hydraulic Conductivity of Unsaturated Soils1, Soil Sci. Soc. Am. J., 44, 892-898, 1980.

Van Genuchten, M. T., Leij, F., and Yates, S.: The RETC code for quantifying the hydraulic functions of unsaturated soils, 1991.

van Mansvelt, J. D., and van der Lubbe, M. J.: Checklist for Sustainable Landscape Management: Final Report of the EU Concerted Action AIR3-CT93-1210, Elsevier, 1998.

Veldkamp, A., and Lambin, E. F.: Predicting land-use change, Agriculture, Ecosystems & Environment, 85, 1-6, 10.1016/s0167-8809(01)00199-2, 2001.

Venkatesh, B., Lakshman, N., Purandara, B., and Reddy, V.: Analysis of observed soil moisture patterns under different land covers in Western Ghats, India, Journal of hydrology, 397, 281-294, 2011.

Verbesselt, J., Hyndman, R., Newnham, G., and Culvenor, D.: Detecting trend and seasonal changes in satellite image time series, Remote sensing of Environment, 114, 106-115, 2010a.

Verbesselt, J., Hyndman, R., Zeileis, A., and Culvenor, D.: Phenological change detection while accounting for abrupt and gradual trends in satellite image time series, Remote Sensing of Environment, 114, 2970-2980, 2010b.

Verburg, P. H., van Eck, J. R. R., de Nijs, T. C. M., Dijst, M. J., and Schot, P.: Determinants of land-use change patterns in the Netherlands, Environment and Planning B: Planning and Design, 31, 125-150, 2004.

Verdin, J., Funk, C., Senay, G., and Choularton, R.: Climate science and famine early warning, Philosophical Transactions of the Royal Society B: Biological Sciences, 360, 2155-2168, 2005.

Vereecken, H., Diels, J., Van Orshoven, J., Feyen, J., and Bouma, J.: Functional evaluation of pedotransfer functions for the estimation of soil hydraulic properties, Soil Science Society of America Journal, 56, 1371-1378, 1992.

Vereecken, H., and Herbst, M.: Statistical regression, Developments in Soil Science, 30, 3-19, 2004.

Vereecken, H., Huisman, J., Bogena, H., Vanderborght, J., Vrugt, J., and Hopmans, J.: On the value of soil moisture measurements in vadose zone hydrology: A review, Water resources research, 44, 2008.

Vereecken, H., Kamai, T., Harter, T., Kasteel, R., Hopmans, J., and Vanderborght, J.: Explaining soil moisture variability as a function of mean soil moisture: A stochastic unsaturated flow perspective, Geophysical Research Letters, 34, 2007.

Vereecken, H., Kasteel, R., Vanderborght, J., and Harter, T.: Upscaling hydraulic properties and soil water flow processes in heterogeneous soils, Vadose Zone Journal, 6, 1-28, 2007.

Vereecken, H., Maes, J., and Feyen, J.: Estimating unsaturated hydraulic conductivity from easily measured soil properties, Soil Science, 149, 1-12, 1990.

Vereecken, H., Maes, J., Feyen, J., and Darius, P.: Estimating the soil moisture retention characteristic from texture, bulk density, and carbon content, Soil Science, 148, 389-403, 1989.

Vereecken, H., Weynants, M., Javaux, M., Pachepsky, Y., Schaap, M., and Van Genuchten, M.: Using pedotransfer functions to estimate the van Genuchten-Mualem soil hydraulic properties: A review, Vadose Zone Journal, 9, 795-820, 2010.

Verhoef, A., van den Hurk, B. J., Jacobs, A. F., and Heusinkveld, B. G.: Thermal soil properties for vineyard (EFEDA-I) and savanna (HAPEX-Sahel) sites, Agricultural and Forest Meteorology, 78, 1-18, 1996.

Veron, S., Paruelo, J., and Oesterheld, M.: Assessing desertification, Journal of Arid Environments, 66, 751-763, 2006.

Versace, V., Ierodiaconou, D., Stagnitti, F., and Hamilton, A.: Appraisal of random and systematic land cover transitions for regional water balance and revegetation strategies, Agriculture, Ecosystems & Environment, 123, 328-336, 2008.

Vicente-Serrano, S. M., Pérez-Cabello, F., and Lasanta, T.: Assessment of radiometric correction techniques in analyzing vegetation variability and change using time series of Landsat images, Remote Sensing of Environment, 112, 3916-3934, 2008.

Villarini, G., Smith, J. A., Baeck, M. L., Vitolo, R., Stephenson, D. B., and Krajewski, W. F.: On the frequency of heavy rainfall for the Midwest of the United States, Journal of Hydrology, 400, 103-120, 2011.

Vogelmann, J. E., Helder, D., Morfitt, R., Choate, M. J., Merchant, J. W., and Bulley, H.: Effects of Landsat 5 Thematic Mapper and Landsat 7 Enhanced Thematic Mapper Plus radiometric and geometric calibrations and corrections on landscape characterization, Remote sensing of environment, 78, 55-70, 2001.

Vogt, J., Safriel, U., Von Maltitz, G., Sokona, Y., Zougmore, R., Bastin, G., and Hill, J.: Monitoring and assessment of land degradation and desertification: Towards new conceptual and integrated approaches, Land Degradation & Development, 22, 150-165, 2011.

Vuichard, N., Ciais, P., Belelli, L., Smith, P., and Valentini, R.: Carbon sequestration due to the abandonment of agriculture in the former USSR since 1990, Global Biogeochemical Cycles, 22, 2008.

Wagenseil, H., and Samimi, C.: Assessing spatio-temporal variations in plant phenology using Fourier analysis on NDVI time series: results from a dry savannah environment in Namibia, International Journal of Remote Sensing, 27, 3455-3471, 2006.

Wahren, A., Feger, K.-H., Schwärzel, K., and Münch, A.: Land-use effects on flood generation–considering soil hydraulic measurements in modelling, Advances in Geosciences, 21, 99-107, 2009.

Wang, S., Fu, B., Gao, G., Yao, X., and Zhou, J.: Soil moisture and evapotranspiration of different land cover types in the Loess Plateau, China, Hydrology and Earth System Sciences, 16, 2883-2892, 2012.

Wang, X. L., and Swail, V. R.: Changes of extreme wave heights in Northern Hemisphere oceans and related atmospheric circulation regimes, Journal of Climate, 14, 2204-2221, 2001.

Ward, J. V., and Stanford, J.: Ecological connectivity in alluvial river ecosystems and its disruption by flow regulation, Regulated Rivers: Research & Management, 11, 105-119, 1995.

Wayne Polley, H., Johnson, H. B., and Derner, J. D.: Soil-and plant-water dynamics in a C3/C4 grassland exposed to a subambient to superambient CO_2 gradient, Global Change Biology, 8, 1118-1129, 2002.

Wessels, K., Prince, S., Frost, P., and Van Zyl, D.: Assessing the effects of human-induced land degradation in the former homelands of northern South Africa with a 1 km AVHRR NDVI time-series, Remote Sensing of Environment, 91, 47-67, 2004.

Wessels, K., Prince, S., Malherbe, J., Small, J., Frost, P., and VanZyl, D.: Can human-induced land degradation be distinguished from the effects of rainfall variability? A case study in South Africa, Journal of Arid Environments, 68, 271-297, 2007.

Western, A. W., Zhou, S.-L., Grayson, R. B., McMahon, T. A., Blöschl, G., and Wilson, D. J.: Spatial correlation of soil moisture in small catchments and its relationship to dominant spatial hydrological processes, Journal of Hydrology, 286, 113-134, 2004.

White, M., Running, S., and Thornton, P.: The impact of growing-season length variability on carbon assimilation and evapotranspiration over 88 years in the eastern US deciduous forest, International Journal of Biometeorology, 42, 139-145, 1999.

White, M. A., Beurs, D., Kirsten, M., Didan, K., Inouye, D. W., Richardsen, A. D., Jensen, O. P., O'Keefe, J., Zhang, G., and Nemani, R. R.: Intercomparison, interpretation, and assessment of spring phenology in North America estimated from remote sensing for 1982–2006, Global Change Biology, 15, 2335-2359, 2009.

Wielicki, B. A., Wong, T., Allan, R. P., Slingo, A., Kiehl, J. T., Soden, B. J., Gordon, C., Miller, A. J., Yang, S.-K., and Randall, D. A.: Evidence for large decadal variability in the tropical mean radiative energy budget, Science, 295, 841-844, 2002.

Wijngaard, J., Klein Tank, A., and Können, G.: Homogeneity of 20th century European daily temperature and precipitation series, International Journal of Climatology, 23, 679-692, 2003.

Wilcox, B. P., Rawls, W., Brakensiek, D., and Wight, J. R.: Predicting runoff from rangeland catchments: A comparison of two models, Water Resources Research, 26, 2401-2410, 1990.

Wilks, D. S.: Statistical methods in the atmospheric sciences, Academic press, 2011.

Wood, A.: Wetland drainage and management in south-west Ethiopia: some environmental experiences of an NGO, The Sahel Workshop, 1996, 119-136,

Wood, A.: Wetlands, gender and poverty: some elements in the development of sustainable and equitable wetland management, Wetlands of Ethiopia, 58, 2003.

Wösten, J.: Pedotransfer functions to evaluate soil quality, Developments in Soil Science, 25, 221-245, 1997.

Wösten, J.: The HYPRES database of hydraulic properties of European soils, Advances in GeoEcology, 135-143, 2000.

Wösten, J., Finke, P., and Jansen, M.: Comparison of class and continuous pedotransfer functions to generate soil hydraulic characteristics, Geoderma, 66, 227-237, 1995.

Wösten, J., Lilly, A., Nemes, A., and Le Bas, C.: Development and use of a database of hydraulic properties of European soils, Geoderma, 90, 169-185, 1999.

Wösten, J., Pachepsky, Y. A., and Rawls, W.: Pedotransfer functions: bridging the gap between available basic soil data and missing soil hydraulic characteristics, Journal of hydrology, 251, 123-150, 2001.

Wu, J., Jelinski, D. E., Luck, M., and Tueller, P. T.: Multiscale analysis of landscape heterogeneity: scale variance and pattern metrics, Geographic Information Sciences, 6, 6-19, 2000.

Wylie, D., Jackson, D. L., Menzel, W. P., and Bates, J. J.: Trends in global cloud cover in two decades of HIRS observations, Journal of climate, 18, 3021-3031, 2005.

233

Wyman, M. S., and Stein, T. V.: Modeling social and land-use/land-cover change data to assess drivers of smallholder deforestation in Belize, Applied Geography, 30, 329-342, 10.1016/j.apgeog.2009.10.001, 2010.

Xiao, X., Boles, S., Liu, J., Zhuang, D., Frolking, S., Li, C., Salas, W., and Moore III, B.: Mapping paddy rice agriculture in southern China using multi-temporal MODIS images, Remote Sensing of Environment, 95, 480-492, 2005.

Xin, Z., Xu, J., and Zheng, W.: Spatiotemporal variations of vegetation cover on the Chinese Loess Plateau (1981–2006): Impacts of climate changes and human activities, Science in China Series D: Earth Sciences, 51, 67-78, 2008.

Yang, L., Wei, W., Chen, L., and Mo, B.: Response of deep soil moisture to land use and afforestation in the semi-arid Loess Plateau, China, Journal of Hydrology, 475, 111-122, 2012.

Yimer, F., Ledin, S., and Abdelkadir, A.: Changes in soil organic carbon and total nitrogen contents in three adjacent land use types in the Bale Mountains, south-eastern highlands of Ethiopia, Forest Ecology and Management, 242, 337-342, 2007.

Yimer, F., Messing, I., Ledin, S., and Abdelkadir, A.: Effects of different land use types on infiltration capacity in a catchment in the highlands of Ethiopia, Soil use and management, 24, 344-349, 2008.

Yoseph, G., and Tadesse, M.: Ethiopian Highlands Reclamation Study: Present and Projected Population, Ministry of Agriculture and Food and Agriculture Organization of the United Nations Addis Ababa, EthiopiaWorking Paper 8, 1984.

Young, M., Gowing, J., Hatibu, N., Mahoo, H., and Payton, R.: Assessment and development of pedotransfer functions for semi-arid sub-Saharan Africa, Physics and Chemistry of the Earth, Part B: Hydrology, Oceans and Atmosphere, 24, 845-849, 1999.

Yu, B., Stott, P., Di, X. Y., and Yu, H. X.: Assessment of land cover changes and their effect on soil organic carbon and soil total nitrogen in Daqing Prefecture, China Land Degradation & Development, DOI: 10.1002/ldr.2169, 10.1002/ldr.2169, 2012.

Yu, G., Song, X., Wang, Q., Liu, Y., Guan, D., Yan, J., Sun, X., Zhang, L., and Wen, X.: Water-use efficiency of forest ecosystems in eastern China and its relations to climatic variables, New Phytologist, 177, 927-937, 2008.

Yuan, D., and Elvidge, C.: NALC land cover change detection pilot study: Washington DC area experiments, Remote sensing of environment, 66, 166-178, 1998.

Zeleke, G., and Hurni, H.: Implications of Land Use and Land Cover Dynamics for Mountain Resource Degradation in the Northwestern Ethiopian Highlands, Mountain Research and Development, 21, 184-191, 10.1659/0276-4741(2001)021[0184:iolual]2.0.co;2, 2001.

Zhang, J. J., Fu, M. C., Zeng, H., Geng, Y. H., and Hassani, F. P.: Variations in ecosystem service values and local economy in response to land use: a case study of Wu'an, China Land Degradation & Development, 24, 236-249. DOI: 210.1002/ldr.1120, 10.1002/ldr.1120, 2013.

Zhang, K., Kimball, J. S., Nemani, R. R., and Running, S. W.: A continuous satellite-derived global record of land surface evapotranspiration from 1983 to 2006, Water Resources Research, 46, 2010.

Zhang, M., Yu, G.-R., Zhuang, J., Gentry, R., Fu, Y.-L., Sun, X.-M., Zhang, L.-M., Wen, X.-F., Wang, Q.-F., and Han, S.-J.: Effects of cloudiness change on net ecosystem exchange, light use efficiency, and water use efficiency in typical ecosystems of China, Agricultural and Forest Meteorology, 151, 803-816, 2011a.

Zhang, W., An, S., Xu, Z., Cui, J., and Xu, Q.: The impact of vegetation and soil on runoff regulation in headwater streams on the east Qinghai–Tibet Plateau, China, Catena, 87, 182-189, 2011b.

Zhang, X., Ren, Y., Yin, Z. Y., Lin, Z., and Zheng, D.: Spatial and temporal variation patterns of reference evapotranspiration across the Qinghai-Tibetan Plateau during 1971–2004, Journal of Geophysical Research: Atmospheres (1984–2012), 114, 2009.

Zhang, X., Zhang, L., Zhao, J., Rustomji, P., and Hairsine, P.: Responses of streamflow to changes in climate and land use/cover in the Loess Plateau, China, Water Resources Research, 44, 2008.

Zhao, M., and Running, S. W.: Drought-induced reduction in global terrestrial net primary production from 2000 through 2009, science, 329, 940-943, 2010.

Zhou, L., Tucker, C. J., Kaufmann, R. K., Slayback, D., Shabanov, N. V., and Myneni, R. B.: Variations in northern vegetation activity inferred from satellite data of vegetation index during 1981 to 1999, Journal of Geophysical Research: Atmospheres (1984–2012), 106, 20069-20083, 2001.

Zhou, X., Lin, H., and White, E.: Surface soil hydraulic properties in four soil series under different land uses and their temporal changes, Catena, 73, 180-188, 2008.

Samenvatting

De verandering van landgebruik en bodembedekking (Land Use and land cover change - LULCC) is een proces dat van groot belang is voor onderzoek naar globale en locale veranderingen van het milieu. LULCC draagt bij aan de hoge mate van bodemerosie en landdegradatie in de hooglanden van Ethiopië. LULCC heeft eco-hydrologische effecten op fysieke en hydrologische eigenschappen van de bodem en op de lokale en regionale variabiliteit van het klimaat. Dit proefschrift analyseert LULCC en de links naar de bodemhydrologie, bodemdegradatie en klimaatschommelingen in het bovenstroomse deel van het Blauwe Nijl stroomgebied in Ethiopië.

Het is haalbaar noch realistisch om gedetailleerd lange termijn onderzoek te doen naar LULCC voor het hele Blauwe Nijl stroomgebied. Daarom richt dit proefschrift zich op één van de bovenstroomse sub-stroomgebieden van de Blauwe Nijl: het Jedeb stroomgebied. Voor dit stroomgebied, dat een oppervlakte van 296.6 km2 beslaat, werd de lange termijn LULCC gekwantificeerd en werden de ruimtelijke determinanten van bodembedekking geïdentificeerd voor de periode 1957-2009. Zwart-wit luchtfoto's van 1957 en Landsat beelden van 1972 (MSS), 1986 (TM), 1994 (TM) en 2009 (TM) werden gebruikt voor classificatie van het landgebruik en bodemdekking in tien klassen door geïntegreerd gebruik van remote sensing (RS) en Geografisch Informatie Systeem (GIS) technieken. Dit onderzoek toont aan dat 46% van de land cover in het studiegebied veranderde over de afgelopen 52 jaar. In 20% gaat het om een netto-verandering. In 26% is de verandering toe te schrijven aan een swap-change (in dit geval verandert de locatie van de landcover in het studiegebied door het gelijktijdig toenemen en verdwijnen van de landcover op verschillende plekken). De meest systematische veranderingen in de landcover zijn die van grasland naar gecultiveerd land (14,8%) en die van bos en struikgebieden naar grasland (3,9%). Met ruimtelijk expliciete logistieke regressie modellen werd bepaald dat de locatie van deze systematische overgangen kan worden verklaard door een combinatie van veranderende bereikbaarheid, biofysische factoren en demografische factoren. De gebruikte modelaanpak leidt tot een beter begrip van de processen van LULCC en voor het identificeren van verklarende factoren. Deze factoren kunnen dienen als uitgangspunt voor verdere analyse en als aanknopingspunt in het formuleren van interventies in het beheer van het stroomgebied.

Conversie van wetlands naar landbouwgrond is een vorm van LULCC die veel voorkomt in Ethiopië maar slecht gedocumenteerd is. Daarom kwantificeerde dit onderzoek de afname van wetlands over van de periode 1986-2005 met behulp van satellietbeelden. De resultaten tonen aan dat 607 km2 van seizoensgebonden wetland en 22,4 km2 open water verloren gingen in het studiegebied. De huidige situatie in de wetlands van de Choke Mountains wordt gekenmerkt door verdere degradatie, die dringend vraagt om aandacht voor het behoud van wetlands in de beheersplannen voor het stroomgebied.

De verbeterde transitie matrix voor LULCC die werd opgesteld in deze studie biedt nuttige informatie over de veranderingen in land cover proporties. De matrix laat echter niet zien hoe de samenstelling en de configuratie van het landschap verandert. De veranderingen in het landschap werden daarom in beeld gebracht door *landscape pattern metrics*. Verschillende *landschape pattern metrics* duiden op significantie veranderingen in de compositie en configuratie van het Jedeb landschap sinds 1957. De toename van het aantal patches (NP) van 1621 tot 5179 in het studiegebied en de afname van de mediaan van de patch grootte van 18.3 naar 5.7 ha in de periode 1957-2009 duidt op een toenemende fragmentatie van het landschap.

De toename in de relatieve variabiliteit van de patch grootte van 1743 ha naar 4933 ha suggereert een toename in de heterogeniteit van het landschap, wat nieuwe management uitdagingen met zich meebrengt. In 2009 wordt het Jedeb stroomgebied gedomineerd door enkele typen patches. De contagion index (CONTAG) waarin en alle patch typen worden geaggregeerd is gestegen van 57% naar 65% in de periode 1957-2009. Deze toename van de dominantie van een klein aantal typen patches uit zich ook in de stijging van de largest patch index (LPI) van 42% naar 68% en een daling van de Shanon's Diversity index (SHDI) waarde van 1.47 naar 1.11 tussen 1957 en 2009. De mean shape index (MSI) rangschikt veranderingen in het Jedeb landschap langs een intuïtieve temporele gradiënt van meest complexe tot eenvoudige vormen (verandering van 1.57 naar 1.24). De patches waren het meest onregelmatig en complex in vorm in 1957 en zijn sindsdien steeds minder onregelmatig en eenvoudiger van vorm geworden. Deze resultaten geven aan dat fragmentatie een vereenvoudiging van patchvorm veroorzaakt.

Bodemdegradatie wordt vaak in verband gebracht met LULCC. Dit proefschrift onderzocht de invloed van LULCC op de belangrijkste bodemkwaliteit indicatoren (SQIs) in de Jedeb stroomgebied van de Blauwe Nijl / Abay. Een (2x5) multivariate variantie-analyse (MANOVA) laat zien dat de hoeveelheid organische stof in de bodem significant afhangt van de LULC en van de hoogte boven zeeniveau. Waar LULC heeft een significant invloed heeft op de bulk dichtheid, heeft de hoogte zowel een significante invloed op de bulk dichtheid, zuurgraad en het siltgehalte. Door herbebossing van onvruchtbaar land met eucalyptusbomen kan de hoeveelheid organische stof in de bodem significant toenemen op midden-hoogtes, maar niet in hoog gelegen gebieden. Gronden onder grasland hadden een significant hogere bulk dichtheid dan bodems onder bossen en struiklanden. Dit geeft aan dat de-vegetatie en conversie naar grasland tot compactie van de bodem kan leiden. Samenvattend heeft de historische LULCC in het Jedeb stroomgebied geleid tot een afname van organische stof in de bodem en een toename van compactie van de bodem. Het onderzoek wijst uit dat bodemindicatoren gebruikt kunnen worden om de invloed van landgebruik en de wijze van beheer op de landdegradatie vast te stellen.

LULCC kan leiden tot veranderingen in de bodem fysische eigenschappen en de totale verdamping met een gewijzigde bodemvochtdynamiek tot gevolg. In dit proefschrift werden zowel ge-downscalede en in-situ bodemvocht observaties gebruikt om de invloed van landcover veranderingen op de bodemvochtdynamiek op basin schaal (199.812 km2) en op deelstroomgebied schaal (296,6 km2) te evalueren. Op basis van remote sensing beelden werden significante verschillen vastgesteld tussen de mediaan van het bodemvochtgehalte van een stabiel bebost referentie traject en andere trajecten die door ANOVA analyse werden vastgesteld (p <0,001). Bodems onder stabiele bossen hebben het hoogste bodemvochtgehalte. In-situ waarnemingen in het case studie deelstroomgebied (Jedeb) bevestigde dat de gemiddelde bodemvochtigheid onder onvruchtbaar land en grasland aanzienlijk verschilde van het gemiddelde bodemvochtgehalte van bosbodems (p <0,001). Zo kon worden aangetoond dat veranderingen in de bodembedekking een belangrijke invloed hebben op het bodemvochtregime op verschillende schalen in het stroomgebied van de Blauwe Nijl.

Vaak zijn niet voor de hele bodemvochtretentiecurve de karakteristieken beschikbaar die nodig zijn in hydrologische modellen om landgebruik effecten op de bodem hydraulische eigenschappen te beoordelen. Daarom worden bodemvochtretentiekarakteristieken vaak voorspeld op basis van direct beschikbare gegevens over de bodem met behulp van pedotransferfuncties (PTF). Vanwege de specifieke eigenschappen van tropische hooglandbodems kunnen elders ontwikkelde PTFs niet worden toegepast op de bodems van de bovenstroomse Blauwe Nijl / Abay regio. Aan de hand van elementaire bodem fysische en chemische eigenschappen kunnen bodemvochtretentiekarakteristieken redelijk goed voorspeld worden met een gemiddelde root mean squared verschil (MRMSD) van 0,0349 cm^3cm^- ^{3} en 0,0508 cm^3cm^{-3} voor respectievelijk point PTF's en de continue PTF's. Het verzamelen en integreren van gefragmenteerde beschikbare bodemgegevens stelt wetenschappers en planners in staat om kaarten van het beschikbare water en bodemvochtretentiekarakteristieken op een stroomgebiedschaal te genereren. Deze studie beveelt verder aan om nader onderzoek te doen op basis van een database die groot genoeg is om representatief te zijn voor alle delen van het textuur driehoek.

Seizoensgebonden veranderingen in vegetatie zijn niet verwerkt in de LULCC analyse. Deze veranderingen kunnen echter belangrijk zijn en zelfs groter dan de veranderingen door je jaren heen. Daarom onderzoekt deze thesis de inter-annual en seizoensgebonden trends in land cover in het Upper Blue Nile/ Abay (UBN) basin. Het onderzoek gebruikt Advanced Very High Resolution Radiometer (AVHRR) gebaseerd op Global Inventory, Monitoring, and Modelling Studies (GIMMS) Normalized Difference Vegetation Index (NDVI) voor een ruwe lange termijn vegetatie trend analyse en Moderate-resolution Imaging Spectroradiometer (MODIS)

NDVI data (MOD13Q1) voor fijnere vegetatie trens analyse. Harmonische analyse en non-parametrische trend tests zijn toegepast op zowel de GIMMS NDVI (1981–2006) en de MODIS NDVI (2001-2011) datasets.

Op basis van een robuuste trend schatter (Theil-Sen slope) werd voor het grootste deel van het UBN (~ 77%) een positieve trend in de maandelijkse GIMMS NDVI vastgesteld met een gemiddelde snelheid van 0,0015 NDVI eenheden per jaar (3,77% jaar-1), waarvan 41,15 % van het basin significante toenames vertoont (p <0,05) met een mediaan snelheid van 0,0023 NDVI eenheden per jaar (5,59% jaar-1) tijdens de onderzochte periode. Echter, uit de fijnere schaal (250 m) MODIS-gebaseerde vegetatie trend analyse bleek dat ongeveer 36% van de UBN een significant dalende trend (P <0,05) in de NDVI index heeft voor de periode 2001-2011 met een gemiddelde snelheid van 0,0768 NDVI yr^{-1}. Seizoensgebonden trendanalyse bleek zeer nuttig voor het identificeren van vegetatieveranderingen die niet uit de inter-annual analyse naar voren kwamen.

Het toeschrijven van waargenomen trends in vegetatie aan bepalende klimatologische factoren is cruciaal. Daarom werden tijdreeksanalyses van de netto primaire productie (NPP, hoeveelheid koolstof in de atmosfeer vast) en de efficiëntie van watergebruik (WUE, hoeveelheid koolstof opname per eenheid van het watergebruik) en correlatie analyses uitgevoerd. De resultaten tonen aan dat de bepalende klimatologische factoren van NPP variëren met de vochtigheid: in vochtige zones zijn regenval en temperatuur dominant, in semi-vochtige zones de temperatuur en het vapour pressure deficit (VPD) en in droge zones de bewolking. In de droge subhumide en vochtige zones van het basin correleert de WUE significant en positief met de maximum temperatuur, de potentiële verdamping, en VPD, hoewel geen enkele klimatologische factor was gecorreleerd met WUE in droge gebieden van het basin.

Concluderend laat dit proefschrift zien dat begrip van LULCC van cruciaal belang is om de potentiële effecten van LULCC op landdegradatie, bodemhydrologie, biodiversiteit (versnippering van het landschap) en de variabiliteit van het klimaat beter te kunnen beheren. Verder onderzoek moet feedback loops tussen LULCC en klimaat, de sociale dimensie van LULCC en de bodemhydrologische impacts van LULCC in beeld brengen door laboratorium onderzoek en velexperimenten. Bovendien word verder onderzoek aanbevolen naar klimaatfactoren van NPP en WUE in verschillende agro-ecologische zones van het Upper Blue Nile /Abay basin waarbij gebruik gemaakt wordt van een hogere spatiale resolutie satellietbeelden en veldobservaties.

About the Author

Ermias Teferi Demessie was born on the 13th of March 1979 in Wolieso, central Ethiopia. Although he was born in Wolieso, he grew up in Butajira, Ethiopia, since the age of two. He received his BSc degree in Soil and Water Conservation from Mekelle University, Ethiopia, in July 2002. He joined the then College of Agricultural Technical Vocational and Educational Training in September 2002 and served as a junior instructor for the courses related to 'Agricultural Water Harvesting Technologies'. In March 2004, he joined the then Ministry of Water Resources as Soil and Water Conservation specialist and there worked for four years. He obtained MSc degree in Earth Sciences (Remote Sensing and GIS) from the Addis Ababa University, Ethiopia, in July 2007. The topic of his master thesis was focused on soil erosion modelling and land use and land cover change in the source region of the Upper Blue Nile river. After receiving his MSc degree, he served Mekelle University for one year as a lecturer with teaching and research responsibilities. Since 2009, he is a lecturer with teaching and research reponsibilities in the Center for Environmental and Developmental Studies, Addis Ababa University. His research interests focus on soil and water management, land surface hydrology, and geospatial technology applications. Besides, he is a reviewer of the journals *Hydrology and Earth System Sciences, Environmental Management, Land degradation and development, and Mountain Sciences* and *Applied Remote Sensing*.

In October 2009, Ermias was officially admitted to UNESCO-IHE Institute for Water Education in the Netherlands as a PhD candidate. In May 2011, he was awarded a research grant from Nuffic to conduct his PhD research. During his PhD research, Ermias presented his work at local and international conferences, which includes the Association of American Geographers (AAG), UNESCO-IHE annual PhD seminar, and Building Climate Resilient Green Economy on Mt. Choke Ecosystem conference. He also supervised 3 MSc students at Addis Ababa University, and one MSc student from Haramaya University. This dissertation presents the results of his PhD study which also contains peer-reviewed articles in scientific journals.

Publications:
Peer-reviewed journal articles

- Ermias Teferi., Uhlenbrook, S., & Bewket, W. (2015). Inter-annual and seasonal trends of vegetation condition in the Upper Blue Nile (Abbay) basin: dual scale time series analysis. Earth Syst. Dynam. Discuss., 6, 267-315. doi:10.5194/esdd-6-169-2015.
- Ermias Teferi, Woldeamlak Bewket, Jochen Wenninger, Stefan Uhlenbrook (2013). Understanding recent land use and land cover dynamics in the source region of the Upper Blue Nile, Ethiopia: Spatially explicit statistical modeling of systematic transitions. Agriculture Ecosystem & Environment 165, 98 – 117.
- T. Gebretsadkan, Y.A. Mohamed, G.D. Betrie, P. van der Zaag, Ermias Teferi (2012). Trend Analysis of Runoff and Sediment Fluxes in the Upper Blue Nile Basin: A Comparative Analysis of Statistical, Physically-based Model, and Land use maps. Journal of Hydrology 482, 57-68.
- Ermias Teferi, Stefan Uhlenbrook, Woldeamlak Bewket, Jochen Wenninger and Belay Simane (2010). The use of remote sensing to quantify wetland loss in the Choke Mountain range, Upper Blue Nile basin, Ethiopia. Hydrology and Earth System Sciences 14: 2415-2428.
- Woldeamlak Bewket and Ermias Teferi (2009). Assessment of soil erosion hazard and prioritization for treatment at the watershed level: case study in the Chemoga watershed, Blue Nile basin, Ethiopia. Land Degradation & Development. 20: 609-622.

Under review and in preparation
- Ermias Teferi, Jochen Wenninger, Stefan Uhlenbrook (2015). Monitoring soil moisture in response to land use and land cover changes using Remote Sensing and in-situ observations. International Journal of Applied Earth Observation and GeoInformation (under review)
- Ermias Teferi, Jochen Wenninger, Stefan Uhlenbrook (2015). Predicting soil water retention characteristics in high altitude tropical soils – A case study from the Upper Blue Nile/Abbay, Ethiopia. CATENA (under review)
- Ermias Teferi, Woldeamlak Bewket, Belay Simane, Stefan Uhlenbrook, Jochen Wenninger (2015). Effects of land use and land cover on selected soil quality indictors in the headwater area of the Blue Nile Basin of Ethiopia. Environemtal Monitoring and Assessment (under review)
- Ermias Teferi (2015). Climatic controls of net primary productivity and ecosystem water-use efficiency in the Upper Blue Nile/Abbay basin. (in preparation)

Articles in Proceedings
- Seleshi Yalew, Ermias Teferi, Ann van Griensven, Stefan Uhlenbrook, Marloes Mul, Johannes van der Kwast, Pieter van der Zaag (2012). Land Use Change and Suitability Assessment in the Upper Blue Nile Basin Under Water Resources and Socio-economic Constraints: A Drive Towards a Decision Support System. In: R. Seppelt, A.A. Voinov, S. Lange, D. Bankamp (Eds.)

241

(2012). Proceedings of 2012 International Congress on Environmental Modelling and Software (pp. 2124-2131). Sixth Biennial Meeting, Leipzig, Germany. ISBN: 978-88-9035-742-8.

- Friedrich J. Koch, Ann van Griensven, Stefan Uhlenbrook, Sirak Tekleab, Ermias Teferi (2012). The Effects of Land use Change on Hydrological Responses in the Choke Mountain Range (Ethiopia) - A new Approach Addressing Land Use Dynamics in the Model SWAT. In: R. Seppelt, A.A. Voinov, S. Lange, D. Bankamp (Eds.) (2012). Proceedings of 2012 International Congress on Environmental Modelling and Software (pp. 3022-3029). Sixth Biennial Meeting, Leipzig, Germany. ISBN: 978-88-9035-742-8.
- Ermias Teferi, Stefan Uhlenbrook, Woldeamlak Bewket, Jochen Wenninger and Belay Simane (2010). The use of remote sensing to quantify wetland loss in the Choke Mountain range, Upper Blue Nile basin, Ethiopia. Hydrology and Earth System Sciences Discussions 7: 6243-6284.
- Ermias Teferi, Dagnachew Legesse, Belay Simane and Woldeamlak Bewket (2008). Prioritization of micro-watersheds on the basis of soil erosion risk in the source region of the Blue Nile river using RUSLE model, remote sensing and GIS: case study in the Muga watershed. In: Hagos, F., Kassie, M., Woldegiorgis, T., Mohammednur, Y., Gebreegziabher, Z. (eds.). Proceedings of collaborative national workshop on sustainable land management research and institutionalization of future collaborative research, (pp. 47-62). Ethiopian Development Research Institute, Addis Ababa.
- Ermias Teferi, Dagnachew Legesse, Belay Simane and Woldeamlak Bewket (2008). Land-use and land-cover dynamics in the source region of the Blue Nile River: case study in the Muga watershed. Proceedings of National Nile Basin Development Forum 2008, March 20-21, 2008, Addis Ababa, Ethiopia.

Book chapter
- Ermias Teferi (2011). The changing bio-physical environments of Mt.Choke In: Simane, B. (Ed.), Building climate resilient green economy in Mt. Choke. Addis Ababa University Printing Press, Debre Markos, p. 43 & 44.

Abstracts
- Ermias Teferi, Woldeamlak Bewket, Stefan Uhlenbrook, Jochen Wenninger. Land use/cover dynamics and landscape pattern analysis (1957-2009) in the source region of the Upper Blue Nile, Ethiopia. 2013 AAG Annual Meeting, Los Angeles, California
- Seleshi Yalew, Pieter van der Zaag, Marloes Mul, Stefan Uhlenbrook, Ermias Teferi, Ann van Griensven, and Johannes van der Kwast (2013). Coupled hydrologic and land use change models for decision making on land and water resources in the Upper Blue Nile basin. Geophysical Research Abstracts Vol. 15, 2013, EGU General Assembly 2013.

T - #0393 - 101024 - C10 - 240/170/15 - PB - 9781138028746 - Gloss Lamination